Starting with Chickens

A Beginner's Guide

Katie Thear

Broad Leys Publishing Ltd

Starting with Chickens

Published by Broad Leys Publishing Ltd.

A catalogue record for this book is available from the British Library.

ISBN: 978 0906137 277

Outside front cover: Speckled Sussex hen. (Katie Thear)

Outside back cover: Abigail, the author's grandaughter. (Katie Thear)

Unless otherwise indicated and credited on the pages,
the illustrations in the book are the author's.

This book is dedicated to the many poultry keepers who have contributed their experiences of this fascinating hobby.

For details of other publications please see page 96.

Broad Leys Publishing Ltd
1 Tenterfields,
Newport, Saffron Walden,
Essex CB11 3UW, UK.
Tel/Fax: 01799 541065
E-mail: kdthear@btinternet.com
Website: www.blpbooks.co.uk

Contents

There are no restrictions on keeping a small flock of chickens, unless there are local by-laws or clauses in property deeds or tenancy agreements.

Bantams are popular in small gardens, with children, and as broody hens.

Introduction

Higgledy Piggledy, my black hen, she lays eggs for gentlemen,
Sometimes nine and sometimes ten, Higgledy Piggledy, my black hen.
Traditional Nursery Rhyme)

This book is written for those who are new to chickens. It gives an over view of the subject, as well as covering all the basic, practical questions which need to be addressed. As it is up-to-date in its range and coverage, I hope that experienced poultry-keepers will also find it useful.

A frequent question is whether there are any restrictions on keeping chickens. Happily, there are none for small flocks, unless there happens to be a specific clause banning them in property title deeds or tenancy agreements. (In some cases, even if such a clause exists, it may have been inserted so long ago that it is no longer being enforced). Those living in urban areas should also check any local by-laws. Ask the local authority!

There is no requirement for the small poultry-keeper to register a flock or test for salmonella, as there was a few years ago. Nor does a small, moveable house require planning permission. It would need to be a large, fixed structure before the local authority would object.

Chickens will live happily in a garden or orchard, which are ideal habitats as long as there is wind shelter and security from predators. They are relatively inexpensive to keep, and if managed sensibly, will not damage plants. Even the smallest garden has room for a few bantams.

Provide a clean house and run, with no stale food and droppings left lying around to attract rodents. Avoid having a cockerel if you have close neighbours who are likely to complain. It is untrue that hens lay better with one. The opposite is the case for they will not be at risk from his spurs or have infections passed to them. Some cockerels can also be dangerous to small children. Game breeds were, after all, valued for their 'fight to the death' qualities before cock fighting was made illegal a century and a half ago! Males are only needed for breeding.

A clean, regularly replenished feeder and drinker are essential. Chickens need a proprietary feed such as layer's pellets or mash (powder form) and a grain feed such as wheat. Clean fresh water is essential at all times. In winter when it is frosty, checks are needed to ensure this.

Flower beds do need to be protected against the hens' scratching, of course. Left unrestricted, they will soon trash a garden. Bedding plants are just like weeds to chickens and are scratched up accordingly. They are no respecters of vegetable beds either, and are partial to greens. In fact, it is a good idea to hang up some garden greens such as cabbages, lettuce and home-grown parsley for them to peck at in the run. They provide added minerals in their diet, and also prevent boredom.

Are Chickens for You?

PROS

There are no restrictions on keeping a small domestic flock of poultry.

There is no requirement to register or test a small flock.

You do not need planning permission for a small, moveable house.

Chickens are relatively cheap to buy.

They are relatively cheap to look after.

Housing needs are fairly modest.

Good range of poultry house manufacturers available.

They are generally quiet (as long as there is no cockerel). The hen will announce to the world that she has just laid an egg, but this lasts only a short time and is usually at a reasonable time, not the crack of dawn.

Vaccinated and salmonella-tested birds are available from reputable suppliers.

They lay fresh eggs for the family.

Surplus eggs can be sold.

They can be reared for the table, if the appropriate breed is kept.

They are interesting in their own right and can prove to be good pets.

They have an important educational role where there are children.

Their scratching activities can be put to use in garden bed preparation, by scratching up weeds and clearing a site.

They are good garden pest clearers.

Their droppings are an excellent source of fertility for the soil.

Specialist poultry feed suppliers have a good range of additive-free feeds.

CONS

If you have 250 or more breeding birds, you must register the flock and test them regularly for salmonella.

There may be clauses restricting the keeping of poultry on your site. Check your deeds!

Unless restricted, they will destroy your flower and vegetable beds.

Hybrid birds are cheap but pure breeds can be expensive.

They may not be available locally, and must either be collected or high delivery costs paid.

Neighbours are likely to complain if they make a noise (cock), smell or attract rats.

Hybrids lay well but many pure breeds do not (unless they happen to be from a good, utility strain).

Many pure breeds stop laying in winter, unless they are provided with extra light.

Poultry houses may not be available locally, but need to be bought by mail.

Some birds offered for sale may not have been vaccinated and salmonella-tested, and you could be introducing problems. Check!

They are vulnerable to predators - foxes, dogs, feral cats, mink and crows or hawks in some areas.

Although they can be left in a secure unit, with food and water for a couple of days, longer absences require a helper to go in and feed them.

Poultry feeds and general supplies may not be available locally.

Not many vets understand chickens.

At some point the birds may need to be put down.

Some people can't abide feathers or feathery creatures.

What do Chickens Need?

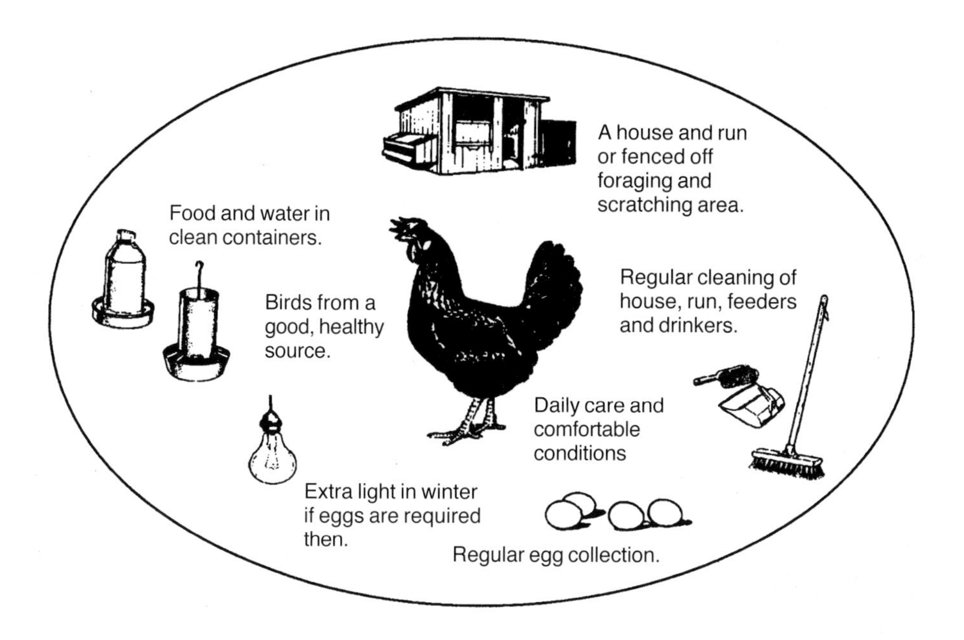

It is also a good idea to make crushed oystershell and poultry grit available. These can be put in a container for the birds to peck at as required, or a handful given with the grain. The crushed oystershell ensures adequate levels of calcium for strong egg shells, while the grit keeps the digestive system working properly. (It helps to break down the grain). They are unlikely to take more than they need.

Like all living creatures, chickens need to be looked after and well cared for if they are to thrive. Weekends away are usually no problem if they are in a self-contained house and run unit, secure from predators. Foxes are the arch-enemy and the birds must be protected against them. Dogs, strange cats and even mink, in some areas, can be a problem. Hawks can also be a problem in some regions.

Large drinkers and feeders will usually last for a couple of days without supervision. Longer periods will require family, friends or neighbours to go in and feed and water the chickens for you. This is often a popular activity for those with children, particularly if they are allowed to take the eggs home.

If you are still unsure as to whether chickens are for you, look at the *Pros and Cons* chart opposite and then decide!

(Katie Thear)

A free-range flock of Bovans Goldline with a home-made, moveable house on wheels. The flock is confined to a field and protected from foxes by electric netting seen in the background. Trees provide wind shelter and a sense of security for the birds.

It's best not to have a cockerel if you have close neighbours. This is a young Friesian.

History of the Chicken

Coin of Carystus.
490 BC

Recent DNA tests have shown that the ancestor of the domestic chicken is a sub-species of the Red Jungle Fowl of Asia, *Gallus gallus gallus,* one of several breeds of Jungle Fowl that still live in the region. Darwin had it nearly right, but he plumped for its cousin *Gallus bankiva* instead. Other Victorian writers such as Temminck had suggested that there was a 'missing link' (calling it *Gallus giganteus)* that would explain the tall upright, tight-feathered birds such as the Asil, but so far, nothing has emerged to support this. The same research indicates that the domestication of chickens occurred far earlier than was previously thought, going back at least 8,000 years. No doubt, more will be revealed as research continues and much that has been written in the past about the history of domestic fowl will need to be reappraised.

From Asia they spread to different areas of the ancient world, where the main interest was in cock fighting. All the ancient cultures enjoyed this pastime, and there are references to it in Indian, Persian, Chinese, Greek and Roman sources. There is an illustration of a cock in the tomb of Tutankhamun in the Valley of Kings, while a cock-crow reminded Peter of his betrayal of Christ. These factors indicate that Egypt and the Holy Land both had their contingent of chickens, although there is no reference to them in the Old Testament and most Egyptian artefacts depict waterfowl and game birds.

There appear to have been two main routes for the distribution of fowl from what is now Thailand. The northern one was via China, Central Asia, Russia and the Near East. The southern route was via India to Persia, Greece and the Mediterranean. It is likely that the conquest of India by Cyrus in 537 BC brought them to Persia. The Greeks called them Persian fowl after Alexander the Great defeated the Persians. Chickens even gave their name to the island of Gallinaria off the coast of Liguria because there were apparently so many of them there. The Phoenicians, who were active traders, probably introduced them to Britain, for Julius Caesar reports their presence when he arrived in 55 BC. Here, too, the main interest was in cock fighting: *"They think it wrong to eat hares, chickens or geese, keeping these creatures only for pleasure and amusement."* The chickens in question were probably the ancestors of the Old English Game, a slight misnomer when one considers that the English tribes did not arrive in any great numbers in these islands until 400 years later. The Romans who had absorbed much of the Greek culture, including the practice of fattening chickens for the table, are thought to have introduced bigger birds that might be the ancestors of the present Dorking. Columella, writing in AD 47, describes 5-toed chickens in Rome, a feature that the Dorking also possesses: *"Let the breeding hens be of a choice, of robust body, square-framed, large and broad breasted, large heads*

Left: Red Jungle Fowl, *Gallus gallus gallus*, from which the domestic chicken has evolved.

Right: The large powerful Asil which differs in size, stance and feathering. (*Harrison Weir, 1896*).

Persian Fowl

Left: Old English Game whose ancestors were probably introduced to Britain by the trading Phoenicians. *(Harrison Weir, 1874)*

Paduan (Poland), one of the crested fowl known for centuries in Europe. *(Androvandi, 16th C. University of Bologna).*

Left: Old type of coloured Dorking whose ancestors were probably introduced to Britain by the Romans. *(G.S. 19th C).*

Left: The Hamburgh whose forbears, like many Northern fowl, were possibly brought by the Saxons and Vikings. *(G.S. 19th C).*

Right: Tall Asiatic fowl introduced in the 19th century brought the gene for brown egg shells and boosted the popularity of poultry keeping. *(Buff Shangai cock. Wingfield & Johnson, 1853).*

(Old illustrations computer-enhanced by Broad Leys Publishing Ltd)

with small erect combs and white ears - those hens are reckoned the purest breed which are five-clawed."

Other immigrants to these islands brought fowl of various kinds. The Vikings and possibly even the Saxons brought the ancestors of light, pheasant-type breeds such as the Hamburgh, Old English Pheasant Fowl and Derbyshire Redcap. However, it was the light Mediterranean breeds such as the Leghorn, Minorca and Spanish, that introduced more prolific egg production, although the eggs were white. Interestingly, the factor for white shells is associated with white ear lobes.

The genetic factor for brown shells was made available by the large Asiatic breeds such as the Shanghai (Langshan), Brahma and Cochin when they were introduced in the 19th century. Their large size caused a great stir and their appearance provided a boost for poultry keeping and showing at the time. It wasn't long before cross-breeding took place to produce a wider range of breeds that were productive and produced brown eggs. The Rhode Island Red from America, for example, was developed from Brown Leghorns and the Asiatic Shanghai, Malay and Java. Another American breed, the Wyandotte (named after a Native American tribe) was developed from Polands and Hamburghs, and the large Asiatic breeds Cochin and Brahma.

After World War 2, there was less commercial interest in the old breeds as hybridisation took place to produce high-yielding layers or rapidly fattening broilers. Most of today's commercial hybrid layers of brown eggs have Rhode Island Red somewhere in their family tree, while the white egg laying hybrids are based on the White Leghorn. It was left largely to the non-commercial sector to keep the traditional breeds going and consequently many of them are no longer as productive as they were, having been bred mainly for showing. Many breeds have become extinct, but in recent years the interest in free-range has stimulated breeding companies to produce more 'traditional type' cross-breeds, and these may in time lead to the standardisation of new breeds. One thing is certain, no-one has produced a fowl equal to that of the old Welsh rhyme:

"Mae gen i iar a cheiliog
A brynais ar ddydd Iau.
Mae'r iar yn dodwy wy bob dydd
A 'r ceiliog yn dodwy dau."

"I have a hen and cockerel
I bought them on a Thursday.
The hen she lays an egg a day
But the cockerel he lays two."

Right: Nest boxes lined with clean wood shavings from untreated wood, or clean chopped straw available from suppliers.

An agricultural show is a good place to see a range of poultry housing and breeds of poultry.

Here, the run extends beneath the house. Note the ramp which has wooden battens to make it easier to climb. The whole structure is moved regularly to fresh ground. The birds are Gold Laced Wyandottes, with the female on the left and the male on the right.

The House and Run

Be it e'er so humble, there's no place like home!

The best time to buy a house is *before* the chickens arrive, but you would be surprised how many people buy chickens on impulse, at a show for example, and then bring them home where there is no house awaiting them.

There are several options. There may be a manufacturer locally who has a ready-made range, or will make a house to order. Perhaps a nearby garden or pet centre sells them. It is also worth checking with pet and livestock feed suppliers for they often act as local agents for national suppliers. We are fortunate in the UK in having several manufacturers who will supply by mail order. The houses may come as flat-packs for self-assembly, or the manufacturer may erect the housing himself. Another option is to make a house and run yourself. There are plans available from a number of sources.

The ideal position is a sunny, well drained area where there is also shade and wind protection. Hens do not like wide open spaces, for they have an instinctive fear of large birds of prey. When one remembers that they are descended from the wild Red Jungle Fowl, this is not surprising. A plane overhead is a bird of prey to a chicken. Trees, shrubs, fences or walls provide a sense of security as well as weather protection, and the average garden usually provides these. Place the house so that the pop-hole (the hens' door) is on the side protected from the prevailing wind. If there is still a whistling wind funnelling into your garden, consider putting a 'porch' around the pop-hole, or place some sort of screen, such as garden mesh, straw bales or wattle hurdles to deflect the wind.

Houses are available in different sizes, so it is important to get one that is appropriate for the number of birds likely to be kept. Ideally, this is a maximum of 10 birds per one square metre or 15kg liveweight of birds per one square metre of floor space. Remember that if you have large breeds such as Brahmas, the normal nest boxes and pop-holes may be too small for them. It is something to mention before ordering, for some manufacturers will adapt their designs and build to order.

Checklist

Check the timber Has it has been treated so that it will stand up to the weather? Pressure treatment is the most effective because it ensures the maximum degree of penetration. If you are re-proofing a house at any time, remember to buy proofer that is *non-toxic to bats.* (Bats are protected and builders are required to use proofers in the roof timbers that will not harm them. If it's safe for bats, it's alright for chickens!)

Support timbers. These are normally 2½-3 cm thick. If too thin the structure may not be strong enough. If too thick, it may be difficult to move.

• *Is the roof sound?* The roof should have an overhang for shedding water. It may be a pitched roof, coming to a point at the top, or be angled away from the door so that water is shed backwards. It may be wood covered with bitumenised felt or be made of a modern material such as *Onduline.* The latter is popular because it does not provide hiding places for mites which can take up residence in the roof felt. Whatever the construction, it needs to be drip-proof and have a good level of insulation.

• *How easy is it to move?* If the house is a moveable one, how is this accomplished? Some houses have wheels, while others may have them offered as an optional extra. The structure needs to be easy to grasp if it is relatively small, so carrying handles need to be provided in the appropriate positions. A larger house may have skids rather than wheels so that it can be pulled. If the house is to be moved by one person, it needs to be easily accomplished.

• *Easy access to the inside.* Does the house have a poultry keeper's door or is there a 'lift-up' section of the roof. Whatever it is, it needs to provide easy and convenient access to the inside of the house for cleaning, and so on.

• *Floor* This may be solid-boarded which is warm, or slatted which is colder but lets droppings fall through. It could be a rigid metal mesh floor which is more secure against rodents. All metal parts should be galvanised.

• *Is it well ventilated?* Stale air can cause health problems so a house needs to provide fresh air without draughts. Depending on the size of house, ventilation is by a window, roof ridge or ventilation holes covered with galvanised wire mesh. Check that air inlets can be opened and closed easily.

• *Pop-hole.* A 'pop-hole' entrance allows birds in and out of the house. This is normally closed by mean of a sliding shutter or 'drop-down' ramp. It is a useful feature to be able to open or close the pop-hole door from outside a run, otherwise you will need to go into the run. All doors, pop-holes and windows should fasten firmly to exclude predators. A lock may also be appropriate to deter thieves. Pop-holes are around 25cm wide x 30cm. Big birds may need 30cm x 38cm.

• *Perches.* There needs to be a minimum 20 cm of space per bird (30 cm if you have large birds). The width is 4-5cm and slightly rounded at the sides for ease of grasping. A perch should be placed higher than the nest boxes so that the birds are not encouraged to sleep in the latter. If there is more than one perch they need to be arranged in such a way that chickens are not directly below each other. Perches should be easy to remove for cleaning.

• *Is there a droppings board under the perch?* This is useful for catching droppings so that they are easily removed by sliding out the board. Alternatively, use plastic sheeting or a droppings box with mesh cover or slats.

Features of a House and Run

All timbers adequately proofed

Easy to dismantle and clean

Sloping roof sheds water away from door side

Ventilation panel above hens' heads

A run that fits snugly onto the pop-hole side but which can be separated from the house

Any metal parts galvanised against rust.

Covered area at end of run for protection in wet or windy weather for birds, feeder and drinker.

Door into run, or means of opening pop-hole from outside the run. Allows birds to range further afield.

Nest boxes can be opened from outside

Solid or slatted floor clear of the ground

Lockable door

Removeable perches placed higher than nest boxes, with droppings board, plastic sheet or slatted box underneath

A guarantee of at least a year that the house and run are built, assembled and proofed to an acceptable standard.

House equipped with wheels, handles or skids for ease of moving

• *Nest boxes.* Nest boxes need to be low down in the darkest area of the house to discourage egg eating. There should be one nest box for every three birds, with wood shavings or sawdust as a liner. Make sure that shavings are from non-treated wood otherwise they may be toxic. Shavings are preferable to hay or field straw which may harbour mites and become mouldy with disease-causing spores *of Aspergillus fumigatus.* 'Farmer's Lung' can also affect humans. Cleaned, chopped straw available in bags is fine.

It is possible to use nest boxes that slope backwards, with a collecting bay at the back, so hens cannot reach the eggs and be tempted to peck them. Nest boxes should be accessible to the poultry keeper from outside the house.

• *Is the house easy to clean?* Fittings should be easily removed for cleaning and the house itself should also be easy to dismantle. A stiff brush and dustpan are good cleaning tools, but having a droppings board or polythene, as referred to above, saves a lot of time and effort.

Below: Small house with run and open pop-hole. Here, the Speckledy hens have sprigs of home-grown parsley for interest and added minerals.

The Run

A *Littleacres* run that can be dismantled for moving and reassembling, and which can be used for most small houses. Netting can be placed on top for extra security.

Most small houses will have a run, either built on or available as an optional extra from the manufacturer. It is a good idea to buy the house and run from the same manufacturer because they will fit easily together. Some manufacturers have interlocking runs so that you can extend the protected area as required. Alternatively, buy a free-standing one that can be dismantled as needed. You can also make your own or provide a fenced off garden area.

Some houses have runs which extend underneath. This maximises the use of space. If part of the run is covered, it provides protection for the birds when it is raining or very sunny, so that they still have the benefit of being outside. It is also useful to be able to put the feeder and drinker under cover.

Perhaps the most important thing about the hens' ranging area is that it is changed regularly. If chickens are left on the same ground all the time, it becomes denuded of grass, and parasites and disease organisms build up. The ground literally becomes 'sick' and the birds succumb accordingly.

The simplest setup is to have a moveable house and run that is moved regularly to fresh grass. Move it as soon as the grass shows signs of wear. Alternatively, have a house with two runs, letting the hens use one run at a time, so that as one is in use, the other run is 'resting'. Some houses have a pop-hole at each end, which is useful for controlling access to alternate runs.

In winter, it may be more appropriate to have the chickens in a house and run on a concrete base which can be hosed down easily, while allowing the birds access to a sanded area for scratching. A really thick layer of hard, coarse wood chips makes an excellent winter run base when the grass has stopped growing. It also absorbs droppings and can be raked over and added to as required. Remember that wood chips are not the same as shavings or shredded prunings which are softer. Shavings are fine inside but not outside. Wood chippings are hard and allow rainwater to drain through leaving the surface dry.

A moveable house and run is very useful if you want the birds to do your vegetable bed clearing in winter. They scratch up weeds and clear the soil of pests. As one area is cleared, move the unit on to the next one. By the time spring arrives, the ground is ready for a light forking and planting.

Fencing

If chickens are in a garden with a substantial fence around it, and there is a family dog in residence, the fox is unlikely to take a chance and pay a visit. He will try somewhere easier. Having said that, every poultry keeper should be aware that the fox is the prime enemy of chickens, and they need to be protected against him. It is no good relying on the local hunt, for that exists purely as a sport for the participants, not as an effective deterrent or control for foxes. There are only two options for the poultry keeper: a fence that is high enough to stop him getting over, or the use of a device such as an electric fence or electronic deterrent.

Traditional fencing

A smooth wooden fence of the type that is used in many gardens is effective, if it is at least 1.8m (6ft) high. Most foxes would baulk at this, although they might try and dig under it, particularly if there is a 'weak' area such as a slight gap that could be enlarged at the bottom. It is worth checking the base of wooden fences and reinforcing them if necessary. Extra boarding or wire mesh can be used.

A really determined fox *can* get over a six foot fence, and if you are unlucky enough to have such a predator in your neighbourhood, then consider putting an extra 30cm (1ft) extension on the top of the fence. If this is angled outwards at an angle of 45^0, it will stop him, for the combination of height and angle will make it impossible to negotiate.

Electric fencing

In an orchard or field the chickens are much more at risk because these often have hedges through which predators can squeeze, or the fencing is lower. Here, unless a high fence of wire mesh poultry netting can be put up, the best option is to use electric poultry netting which can be erected and moved as necessary. It consists of a series of plastic posts with metal spikes which are hammered in, netting which is electroplastic twine, straining post guys and pegs, and a rechargeable battery unit. Complete packages are available for the small poultry keeper, and represent the best value, rather than trying to put together a system of disparate items yourself

Other devices

There are other devices available. It is possible to use a system such as the German *AXT Electronic* device which automatically closes the pop-hole and opens it in the morning. You need to ensure that the chickens put themselves to bed before the fox starts his prowl. Providing food in their house is an incentive for them to do this, and they will then form the habit of roosting before it gets dark. You need to be aware, of course, that in some areas where there are many foxes, they may go on the prowl *before* dusk arrives.

Here, an electronic device is being used to close the pop-hole automatically and keep out the fox. *(AXT Electronic).*

Moveable electrc netting is an effective way of keeping the fox away from a flock, as well as making pasture available in rotation.The dark hens are Bovans Nera. The light ones are Bovans Goldline.

There are also electronic devices which can be used to form an invisible 'beam' around the area. If broken by an intruder (fox or human) an alarm system is sounded or alternatively (and much more desirable if you have close neighbours), a bright light comes on to deter the prowler.

Products such as *Renardine* are available which are sprayed around the perimeter of the area to be protected, and are said to have a deterring effect. I have never used these and so cannot comment on their effectiveness. They need to be reapplied periodically, particularly after rain which will wash away the effect. Mention was made earlier of the fact that the presence of a dog has a deterring effect on foxes. Many poultry keepers walk their dog around the poultry enclosure in the evening so that the scent is transmitted to any foxes. If you live near a zoo, placing lion or tiger droppings is said to be effective. Llamas and alpacas are also reputed to chase foxes away.

Left: Moveable, electrified poultry netting. The holes are smaller at the bottom than at the top to keep in young birds. *(Flexinet)*

A Range of Small Poultry Houses and Runs

These photographs show a range of poultry houses for different uses and flock sizes. There are also many others. Many of the manufacturers attend agricultural and poultry shows where it is possible to view the buildings.

Trio house for 3 birds. This can be free-standing or used with a run. *(Woodside).*

Huntsford unit for a breeding trio or a hen and chicks. The house has alternate runs. *(Littleacres)*

An easily moved ark for a breeding trio or hen and chicks. *(Walner)*

Boughton unit for up to 6 hens. The roosting area is above the ranging area. *(Forsham)*

Ark with ramp and under-house area, surrounded by a moveable run. *(Lindasgrove)*

Minicoop and run for 6-8 birds. The unit is on skids for moving. (*Gardencraft)*

House and run to which the hens can be confined if necessary. The unit is raised off the ground to deter vermin. *(Smiths Sectional Buildings)*

Here, the same house is shown with the droppings boards pulled out. *(Smiths Sectional Buildings)*

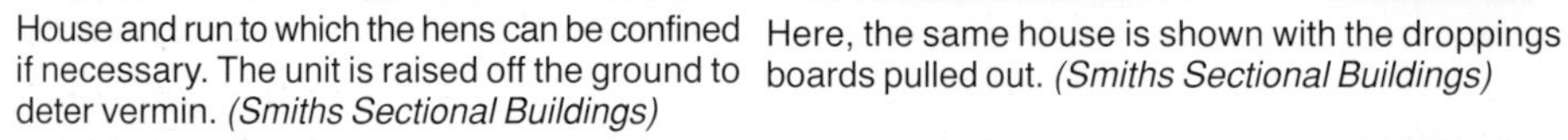

A home-made bantam house and run.

A *Littleacres* house and run with added wheels for moving.

Half-Penthouse from the *Domestic Fowl Trust.* The ramp up to the pop-hole is not shown here.

DIY is an option with appropriate plans and skills. Here, *Onduline* is being sawn to make the roof.

Choosing a Breed

Choice, choice! I want chickens not choice.
(Buyer on being told that there was a three month waiting list for some pure breeds, 1998)

There is a such a wide variety of chickens to choose from that it makes sense to look carefully at the options before making a decision. The factors to take into consideration are whether you want *pure breeds, first crosses, hybrids* or *bantams,* whether they are for *showing, producing* or just *enjoying.*

The bigger breeds are sometimes referred to as *heavy* or *sitting* breeds, and were traditionally used as table birds. The smaller breeds are called *light* breeds. These are generally more productive and flighty, and are used as layers. The most prolific layers were traditionally White Leghorn, Rhode Island Red, Light Sussex, Wyandotte and Barred Plymouth Rock. Now the best layers are hybrids. Some breeds are referred to as *dual-purpose,* because they were used for both table and egg production.

Pure breeds

These are breeds which will breed true, ie, the young will always resemble the parents. Each breed has its own breed club which looks after its interests, and draws up a set of *Standards* for the breed. The best examples of the breed will be those most closely adhering to the *Standards.* All the breed clubs are affiliated to the *Poultry Club of Great Britain* in the UK, or to the *American Poultry Association* in the USA. There are differences in their respective *Standards.*

In recent years, the pure breeds have been bred mainly for show, with the emphasis on appearance rather than on production. They will, therefore, not lay as many eggs as the more modern hybrids. This may not be important to those who just want a few birds in the garden, but it is obviously crucial to those who want to sell eggs from their flock. There are a few flocks of utility pure breeds, where the emphasis has been on production, but these days, there are comparatively few breeders with good, pre-war strains. The main utility breeds are Rhode Island Red, Sussex, Leghorn, Wyandotte, Plymouth Rock, Maran and Welsummer.

There is currently great concern about the declining productivity and fertility of the pure breeds of large fowl, for many have become too in-bred, and there has been too much emphasis on appearance at the expense of productivity. Organisations such as the *Utility Poultry Breeders' Association* are trying to redress matters.

First crosses

When two different pure breeds are crossed, the progeny are called *first crosses,* eg, when a Rhode Island Red male is mated to a Light Sussex female, the young are first crosses, usually referred to as RIR x LS.

Left: Scots Dumpy cock, a short-legged breed that in Scotland has also been called Crawler, Creeper or Bakie. The breed produces white eggs.

Right: Appenzeller Barthuhner, a hardy old Swiss breed that lays white eggs.

The RIR x LS is also sex-linked, ie, the chicks can be identified as male or female when they hatch. In this cross, the males are always silvery while the females are orange-brown.The cross does not work the other way round.

Fancy breeds

These have been bred for the 'fancy' or 'showing' sector. They are pure breeds, of course, but they do not have the productive capacity of the utility breeds. They include Sebrights, Frizzles, Yokohamas, and many others.

Hybrids

These are birds of mixed parents and grandparents which have been developed for production by selecting from the best strains. A *strain* is a family line which has a particularly desirable feature, as far as the breeder is concerned. This might be a good record of eggs laid, good shell quality or excellent feathering. Examples of commercial laying hybrids are Hisex Brown and Babcock Brown 380. All the brown egg laying hybrids are based on the Rhode Island Red while the white egg layers are based on the White Leghorn.

Hybrids are generally the best choice for commercial production. Some have been bred for laying, while others are developed for the table. The latter are usually referred to as broilers. Initially developed for intensive battery and broiler conditions, hybrids have also been bred for free-range conditions. The layers are slightly heavier at point-of-lay so that they are better equipped to cope with outside conditions. Examples include Black Rock, Bovans Nera, Hebden Black, Lohmann Brown, Speckledy, Hisex Ranger, Calder Ranger (originally Columbian Blacktail) and Babcock B380.

The most common, commercial table birds are the white feathered Cobb 500 and Ross broilers. These were bred for intensive conditions and grow quickly, although they do adapt to outside conditions. Table breeds have also been specifically bred for free-range, and these grow more slowly and are white or brown feathered, although grey and other colours are also available, depending on the strain. Free-range table breeds include strains of Hybro, Hubbard-ISA and Sasso. The range of different feather colouring depends on the male used. (See page 50).

Bantams

These are smaller birds than the large fowl. Some are naturally occurring bantams, such as Pekins and Sebrights, for which there are no counterparts amongst large fowl. Others are miniaturised or bred-down versions of large fowl. They all have their breed clubs and there are sets of *Standards* for them. They are popular with children, show enthusiasts and those with smaller gardens, although it should be remembered that bantam cockerels have a particularly shrill call if there are close neighbours. Some bantams are also useful as broody hens for incubating the fertile eggs of other breeds.

Although the fertility and productivity of many large pure breeds have declined, there is evidence that some bantam strains have not been affected to the same degree. For the small, household flock, bantams may therefore be the best choice for supplying the family eggs, as long as they are utility breeds, as referred to earlier.

Hard feather and soft feather breeds

Breeds are sometimes referred to as being *hard* or *soft feathered*. This refers to the relative closeness of the feathers in relation to each other. Hard feather breeds have close-fitting, smooth plumage, while soft feather ones have looser, more fluffy plumage. Game fowl are hard feathered, while other breeds are soft feathered.

Some soft feather breeds which have been developed for show tend to be more 'fluffy' than their utilitarian counterparts, while hybrids for free-range are increasingly being selected for 'close feathering', so there is a growing diversity. The Silkie has feathers lacking barbs so that the plumage is naturally fluffy and silky.

Variation

With so many introductions and inter-breeding, it is inevitable that there is a great variation in colour and patterning of the plumage. The type of comb, wattles and ear lobes vary, while some breeds have crests, muffling (side whiskers) or beards. Legs may be feathered or clean. The general size, build and stance of the birds shows considerable variation, from the tiny Frizzle bantam to the heavyweight Asiatics such as Cochins and Brahmas. The following pages indicate some of these variations, and provide further details of the breeds. Going to see poultry at shows and farm parks is a good idea, while the book *British Poultry Standards* is recommended as a colour reference guide to the recognised pure breeds.

Table of Breeds

Heavy Breeds

Australorp
Barnevelder
Brahma
Bresse
Cochin
Crèvecoeur
Croad Langshan
Dominique
Dorking
Faverolles
Frizzle
German Langshan
Houdan
Ixworth
Jersey Giant
La Flèche
Marans
Modern Langshan
New Hampshire Red
Norfolk Grey
North Holland Blue
Orloff
Orpington
Plymouth Rock
Rhodebar
Rhode Island Red
Sussex
Transylvanian Naked Neck
Wyandotte
Wybar

True Bantams

Belgian Barbu d'Anvers
Belgian Rumpless d'Anvers
Belgian Barbu d'Uccle
Belgian Rumpless d'Uccle
Belgian Barbu de Watermael
Booted
Dutch
Japanese
Nankin
Pekin
Rosecomb
Sebright
Tuzo

Light Breeds

Ancona
Andalusian
Appenzeller
Araucana
Augsburger
Brabanter
Brakel
Breda
Campine
Derbyshire Redcap
Fayoumi
Friesian
Italiener
Hamburgh
Kraienkoppe
Lakenvelder
Legbar
Leghorn
Marsh Daisy
Minorca
Old English Pheasant Fowl
Poland
Rheinlander
Rumpless Araucana
Scots Dumpy
Scots Grey
Sicilian Buttercup
Silkie
Spanish
Sulmtaler
Sultan
Sumatra
Uilebaard
Vorwerk
Welbar
Welsummer
Yokohama

Hard Feathered

Asil
Belgian Game
Carlisle Old English Game
Indian Game
Ko-Shamo
Malay
Modern Game
Nankin Shamo
Oxford Old English Game
Rumpless Game
Shamo
Tuzo
Yamato-Gunkei

Hybrid Layers

Babcock B380
Black Rock
Bovans Goldline
Bovans Nera
Calder Ranger
Hebden Black
Hisex Brown
Hisex Ranger
ISA Brown (Warren)
Lohmann Brown
Lohmann Tradition
Speckledy
White Star (white eggs)

Hybrid Broilers

Cobb 500
Ross White broiler
Hybro White
Hubbard-ISA *
Sasso *

* Sasso and Hubbard-ISA strains of table birds are available under a range of names, depending on the supplier, eg, Devon Bronze, Cotswold White, Poulet Gaulois, etc.

Bantam versions of large fowl are heavy, light, hard or soft-feathered, depending on the specific breed from which they were scaled down.

There are many other breeds of domestic fowl in different parts of the world which are not referred to here. The book *Poultry of the World* by Loyl Stromberg provides a good but rather over-priced coverage of these.

Types of comb *(after Wippell)*

Combs

1. **Rose** (with leader following line of back) - as in Barbu d'Anvers Belgian bantam.
2. **Pea or Triple** - as in the Asil and Sumatra.
3. **Rose** (with level leader) - as in the Hamburgh.
4. **Walnut** - as in the Yokohama.
5. **Cap** - as in the Debyshire Redcap.
6. **Mulberry** - as in the Orloff.
7. **Medium Single** - as in the Australorp.
8. **Large Single** - as in the Minorca.
9. **Cup** - as in the Sicilian Buttercup.
10. **Rose** (long leader) - as in the Old English Pheasant Fowl.
11. **Leaf** - as in the Houdan.
12. **Horn** - as in La Fleche, Poland and Crèvecoeur.
13. **Small Single** - as in the Cochin.
14. **Folded Single** - as in the female Leghorn.
15. **Semi-Erect** - as in the female Dorking.

Types of feathering

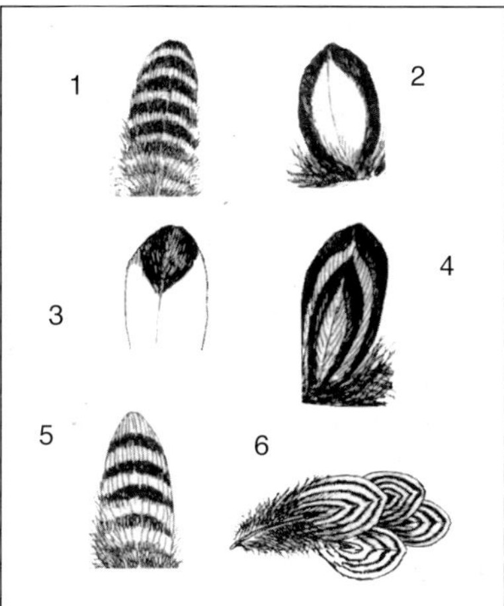

1. **Barring** - where a light and dark colour alternate horizontally across a feather, as in the Plymouth Rock.
2. **Lacing** - an edging of a different colour, as in the Sebright and Andalusian.
3. **Spangling** - an area of a different colour at the end of each feather, as in the Ancona.
4. **Double-lacing** - two lines of a darker colour, as in the Indian Game.
5. **Cuckoo** - barring which is less regular and where stripes can run into each other, as in Cuckoo Marans.
6. **Pencilling** - fine stripes which may follow the shape of the feather, as in the Silver Grey Dorking, or horizontal, as in the Hamburgh.

Ear lobes and egg colour

There is a genetic link between the colour of the ear lobes and that of egg shells. Chickens with white lobes, such as the Spanish, Minorca and the Leghorn, always lay white eggs.

Spanish *(Harrison Weir)*

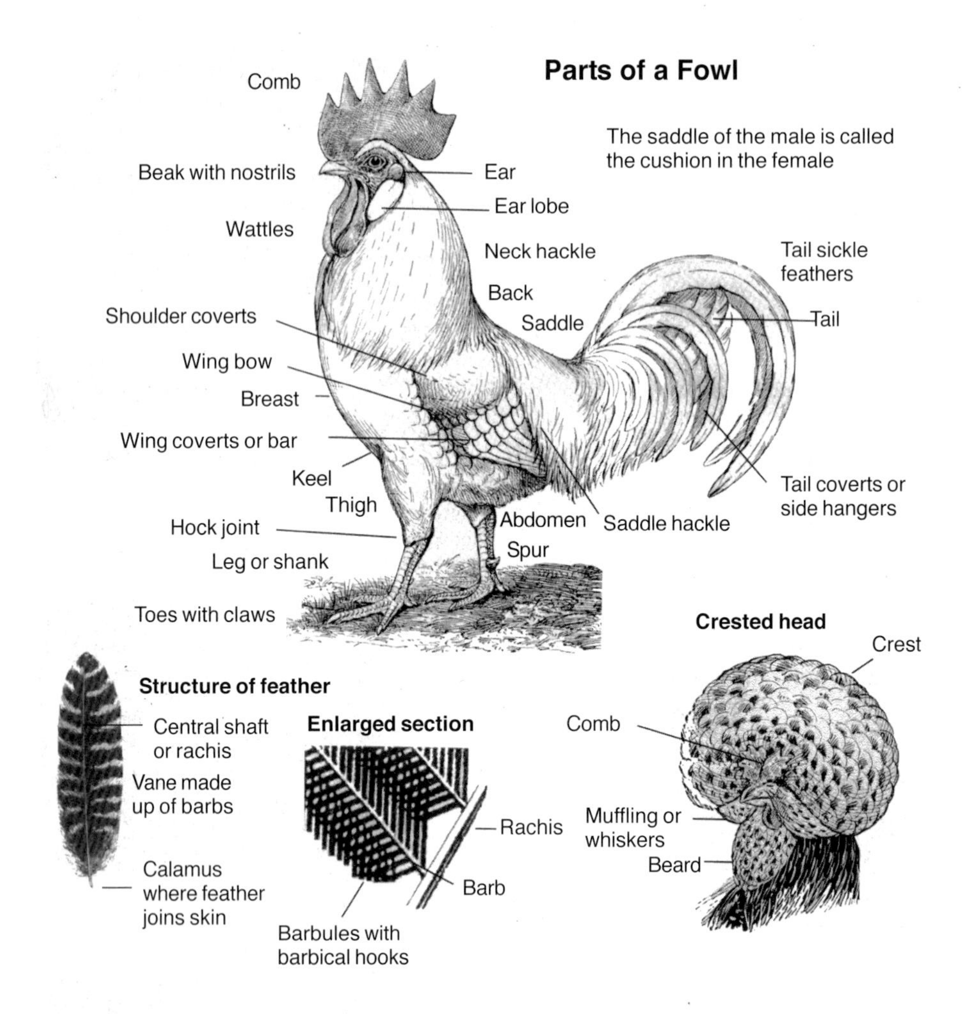

Left: **Pair of Sultans** *(Ludlow)*
A snow-white breed from Turkey, which shows a wide variety of external characteristics differing from the norm: large crest, muffling, beard, five toes and feathered legs with vulture hocks (stiff feathers which project outwards).

Right: **Indian Game** *(Harrison Weir)*
This breed could not be more different, with its tight (hard) feathering, clean legs and small comb.

Description of breeds

This section provides more details of the breeds and their development. It also includes details of the hybrid breeds that have been developed from them for the free-range sector.

The photos do not purport to illustrate show standard birds; they are merely to indicate the diversity of breeds available. Some are the author's birds, while others are taken at farm parks and shows; some are therefore shown in cages.

Ancona

Pure breeds of large fowl

Ancona (see also page 59)
An Italian light breed first introduced into Britain in 1851, the Ancona is related to the Black Leghorn which displaced it commercially. Subsequent development was by fanciers. The plumage is beetle-green with attractive, white V-shaped spangling. It lays white-cream coloured eggs. There is also a bantam. It can be a nervous bird.

Andalusians *(Ludlow)*

Andalusian
Developed from the Spanish breed and introduced into Britain in the nineteenth century, the Andalusian is another striking light breed with its slate-blue plumage and black lacing. Originally, Andalusians were black or white and breeding may produce both colours in a hatch. It is difficult to breed slate-blue birds with well marked lacing. Using dark birds is a starting point. Eggs are white. There is a bantam version.

Silver Appenzeller Spitzhauben

Appenzeller
Named after Appenzelle in Switzerland, this old light breed lays white eggs. It is hardy and is thought to be related to the Northern fowl introduced by the Vikings.

Lavender Araucana

Rumpless Araucana

Asil

The Silver Spangled Appenzeller Spitzhauben has a distinctive, forward-pointing crest, similar to lace bonnets of the region. The plumage is silvery white with black tipping. There is also a Gold Spangled and a Black variety, but no bantam version. The Barthuhner Appenzeller (see page 22) is a different type, with a rose comb rather than a horn comb with crest.

Araucana

Orginating in Chile, this light breed lays blue-green eggs. It has muffling on the face and ears. The first variety standardised in the UK in the 1930s, was the Lavender. There are also Blue, Black-Red, Silver Duckwing, Golden Duckwing, Blue-Red, Pyle, Crele, Spangled, Cuckoo, Black, and White varieties. The bluish-green egg factor is dominant so crossing with other breeds will produce this colour in the eggs of the progeny. Mating birds with excessive crests and muffs is best avoided as it is associated with a lethal gene.There is a bantam version. The Rumpless Araucana lacks the prominent rump and tail feathering of the Araucana. It was introduced to Britain in the 1920s and, for its relatively small size, lays a large egg.

Asil (See also page 10)

A large, hard-feathered bird from Asia this is an ancient breed that was bred for fighting. It lays tinted eggs but the numbers are usually low. The Indian Game is descended from the Asil.

Augsburger

A light breed from Germany, the Augsburger has a distinct cup comb, a legacy from the French breed La Fleche which figures in its development. It has greenish-black plumage and lays white eggs. There is a bantam version.

Australorp

Based on the British Black Orpington, the heavy breed Australorp was bred in Australia using outcrossings of White Leghorn, Minorca and Langshan. It became an important utility breed with record egg production there. In Britain, the Black Orpington was developed for show, losing its original utility qualities along the way. Eggs are tinted to brown. The Australorp was introduced to Britain in the 1920s. It has glossy black plumage. There is a Blue and a bantam version.

Australorp

Autosexing breeds

Autosexing breeds were created in 1929 for identification of chicks at hatch. Breeds such as Barred Plymouth Rock and Maran, which have the barring gene, have chicks with a light head patch. When these breeds are crossed with any brown-feathered breed, the female chicks have the head patch, but the males have them extending down the back. Autosexing breeds were developed which also had this feature, but without having to be crossed with another breed. There are several types and all have bantam forms.

Cambar based on the Campine.

Rhodebar based on the Rhode Island Red.

Legbar developed from the Leghorn. There are Cream, Gold and Silver varieties. The Cream Legbar differs from the Gold and Silver varieties in having a crest and lays blue-green eggs, legacies from the Araucana.

Cream Legbars, one of the autosexing breeds

Barnevelder

A heavy breed from the Netherlands, it was introduced into Britain in the 1920s as a layer of dark brown eggs. It was developed from the Croad Langshan and Gold-Laced Wyandotte. There is a Partridge and an attractive Double-Laced variety which has greenish-black lacing on reddish-brown plumage. The Partridge is similar, but with single lacing. There is a bantam version.

Barnevelder

Brabanter *(Edward Brown).*

Dark Brahma

Silver Brakel *(Edward Brown)*

Silver Campine

Brabanter

The Brabanter is a light breed from the Netherlands. It is distinguished by its cup comb and crest, as well as by its beard and muffling. There is a Silver and a Golden variety, but they are rare.

Brahma

The Brahma was originally developed from large birds with feathered legs which were imported from China into the USA, and crossed with Malay type fowl from India. With other Asiatic breeds, it introduced the factor for brown egg shells. The first introductions were birds presented to Queen Victoria in 1852. Varieties include Dark, Light, White, Gold, Buff, Columbian and bantam.

Brakel

A Belgian bird closely related to the Campine, the Brakel is a light breed, first introduced into Britain at the end of the last century. It produces white eggs. There are two varieties; the Silver and Golden. There is also a bantam.

Breda

A Dutch light breed, the Breda is unique in not having a comb, although some of the head feathers are upright. It lays tinted eggs and has feathered legs. Varieties include Black, White, Blue, Laced Blue, Cuckoo and bantam.

Bresse (See page 37)

An ancient French breed, the Bresse is the premier table breed in France, and is protected by an *Appellation Controllé* so that only birds bred in the specified area can be called by that name. Those sold outside the area must be called Gauloise. It is a light breed related to the Castilian, Spanish and La Fleche, but with the ability to gain weight rapidly.

Campine

An ancient light breed and white egg layer from Belgium, the Campine was introduced in the 19th century and played an important role in the development of the autosexing breeds, referred to earlier. It shares a common ancestry with the Brakel. Like many of the light breeds, it can be flighty. There is a bantam.

Buff Cochin cock.

Cochin (see also page 36)
A large, feather-legged Asiatic breed, the Cochin caused a sensation when first introduced in the last century. It stimulated a great interest in poultry keeping and showing. Eggs are tinted. Varieties are Black, Blue, Cuckoo, Partridge, and White. There is no bantam.

Crèvecoeur
This is a rather exotic, heavy old French breed with gleaming black plumage with a greenish hue. There is a distinct crest and a V-shaped comb. Eggs are white. There is a bantam version.

Crevecoeur

Croad Langshan
The Croad Langshan is a heavy, feather-legged breed developed from Asiatic birds by a Major Croad in the last century. They lay brown eggs. There is a Black, and a White variety as well as a bantam.

Derbyshire Redcap
A white egg layer, the Derbyshire Redcap was used for both meat and eggs, although it is classified as a light breed. It needs wide ranging and high fencing. There is no bantam.

Derbyshire Redcap

Dominique
A heavy breed and layer of brown eggs, the Dominique is the oldest breed in the USA, with a good reputation for laying. It is thought to have been developed from Asiatic birds and the Hamburgh. The plumage is slate-blue with barring. There is a bantam.

White form of Dominique in the USA.

Silver Grey Dorking bantam male.

Salmon Faverolles

Fayoumi

Dorking (See also Page 36)
One of the oldest British table breeds, probably introduced by the Romans, the Dorking has five toes rather than the usual four. A heavy breed with tinted eggs, it was often crossed with Indian Game to produce table birds. There are Dark, Red, Cuckoo, Silver Grey, White and bantam varieties.

Faverolles
The Faverolles is a heavy French breed traditionally regarded as a dual-purpose, egg and table breed. Its distinctive feature is its impressive whiskers or muffling. Produced by inputs from Brahma, Houdan, Cochin and Dorking, it was introduced into Britain near the end of the last century. There are Salmon, Buff, Black, Laced-Blue, Cuckoo, Ermine, White and bantam varieties.

Fayoumi
A hardy and elegant, light breed from Egypt, the Fayoumi was introduced into Britain in 1984. The hens lay white or cream eggs. There is a Silver Pencilled and a Gold Pencilled variety. The birds can be flighty and vocal, needing extensive roaming and secure housing. There is a bantam version.

Friesian (See page 8)
A light breed from Holland, the Friesian lays white eggs and is available as Gold Pencilled, Silver Pencilled and Chamois Pencilled. There is a bantam.

Frizzle (see also page opposite)
An Asiatic heavy breed noted for its curling feathers, the Frizzle is an exhibition bird. There is a bantam version which is more common. There is a wide range of feather colouring and patterning. Eggs are white.

Hamburgh (see pages 34 and 40)
The pheasant-like breed has been known for hundreds of years in the North of England.

(Continued on Page 34)

Breeds - large and small, useful and pretty

Buff Orpington male, a large stately British bird that looks well in a garden setting, but although popular, the breed is not very productive.

Frizzle bantam. Originating in Asia, this little bird has strange backward curling plumage and is kept purely for its appearance, for it lays few eggs.

ISA-Brown (originally called Warren) - a highly productive hybrid layer of brown eggs. Here, she is about to drop a feather in the annual moult.

Above: The delightful little Lavender Pekin.

There is a wide range of options for the poultry keeper, from pure breeds to hybrids, and from large fowl to bantams. They include *fancy* birds kept for their appearance and *utility* kept for their production. There is truly a breed for every choice.

Frizzle

Gold Pencilled Hamburgh

Houdan

Once a good layer, it was overshadowed by the introduction of the Leghorn, and is now a show bird. The varieties in which it is found include Silver Spangled, Black, Gold-Pencilled, Silver Pencilled, and Gold Spangled. There is a bantam version for all varieties except Black.

Houdan
This large French, crested breed has five toes rather than four. Its plumage is greenish black with white mottling. One of France's oldest table breeds, it was introduced to Britain in 1850. Eggs are white. There is a bantam form.

Indian Game (See pages 26 and 41)
A heavy, hard-feathered breed, the Indian Game was bred from the Asil, Malay and Old English Game. It lays tinted eggs, although few in number. A wide-breasted bird, it was traditionally used for crossing with Sussex and Dorkings for table bird production. There are Dark, Jubilee, Double-Laced varieties, and a bantam.

Ixworth
A heavy British breed developed in 1932 as a table bird, the all-white Ixworth's makeup includes a variety of breeds, including Sussex, Orpington, Minorca and Indian Game. It produces tinted eggs. There is a bantam version.

Jersey Giant
Bred in USA from Dark Brahma, Langshan and Indian Game, the early Jersey Giants were well-named for they were true heavyweights. Modern specimens are lighter. There are White, Black, Blue and bantam varieties. Eggs are brown.

Jungle Fowl (See pages 10 and 37)
A pheasant-like group of birds found in Asia has provided the genetic pool from which the domestic chicken has evolved.

Kraienköppe
A light breed from Germany and the Netherlands, the Kraienköppe is called the Twentse by the Dutch. Bred from indigenous fowl, Malay and later Silver Duckwing Leghorns, the breed produces white eggs. There is a bantam.

La Flèche (see page 36)
This is a heavy French table breed whose two-spiked comb gives it a devilish look. It is a relative of the Crêvecoeur and has glossy black plumage with a greenish sheen. The eggs are tinted. There is also a bantam version.

German Langshan
A tall bird, originally from China, this is the smooth-legged version of the Langshan breed, and was developed in Germany. Eggs are cream. Varieties are Black, White, Blue and bantam.

Lakenvelder (See page 41)
This attractive light German breed is black and white with an erect comb. Eggs are tinted. There is a bantam.

Leghorn (See pages 36, 38 and 63)
The Mediterranean Leghorn had a major influence on egg production. (All white egg hybrids are based on the White Leghorn). There are also Black, Brown, Blue, Buff, Cuckoo, Golden Duckwing, Silver Duckwing, Exchequer, Mottled, Partridge, and Pyle varieties. The White is the most productive. There are bantam versions.

Malay (see page 38)
An Asian game bird, the Malay was used in the development of the Indian Game. It lays tinted eggs. Varieties include Black, Black-Red, White, Spangled, and Pyle. There is a bantam version.

Ixworth

Jersey Giants

German Langshan bantams.

Pure Breeds

Buff Sussex

Partridge Cochin in his resplendent colours.

A rare Lavender Leghorn hen.

La Fleche, an old French table breed.

Silver Grey Dorking hen

White Crested Polands

Red Jungle Fowl, ancestor of the modern hen.

Canadian utility strain of Barred Plymouth Rock.

La Bresse (Gauloise), an old French table breed.

Sicilian Buttercup,

Young Silver Laced Wyandotte

Miniature Light Sussex

White Leghorns

Malay (Ludlow)

Modern Game *(Ludlow)*

Marans (see opposite and page 40)
A heavy French, dual-purpose breed, Marans can produce dark, speckled eggs, but only if from a good strain. Varieties are Dark Cuckoo, Silver Cuckoo, Black, and Golden. There is also a bantam version.

Marsh Daisy (see page 40)
This is a light and rare, tinted egg layer first created at the end of the last century, using bantam Old English Game, Malay, White Leghorn and Black Hamburgh. There are Brown, Buff, Black, White, and Wheaten, but no bantam.

Minorca
Another of the Mediterranean breeds, it was at one time renowned for its large, white eggs but tended to be over-developed for exaggerated comb and wattles which had a negative effect on production. It was also crossed with heavier birds to develop its size, but this, too, had a detrimental effect on egg production. There are Black, White, Blue and bantams.

Modern Game
The Modern Game was bred in Britain from Malay crosses. All upright in stance, there are numerous varieties, including Black, Black-Red, Blue, White, Blue-Red, Silver-Blue, Lemon-Blue Birchen, Brown-Red, Golden Duckwing, Silver Duckwing, Wheaten, and Pyle. Eggs are tinted.

Modern Langshan
A tall breed, the Modern Langshan is different in shape from the Croad Langshan, and unlike the German Langshan, has feathered shanks. Eggs are brown.

New Hampshire Red (See page 41)
A heavy American breed based on the Rhode Island Red, it produces tinted to brown eggs. There is a bantam version.

Norfolk Grey (see page 42)
Based on the Duckwing Leghorn and game birds, the Norfolk Grey is a heavy breed that was developed in Norfolk in the 1920s. Eggs are tinted. There is a bantam version.

North Holland Blue
Bred in the Netherlands, the North Holland Blue is a barred breed with feathered legs. It lays tinted eggs. There is a bantam form but it is rare.

Old English Game (See pages 40 & 50)
Ancestors of the Old English Game were probably the type introduced by the trading Phoenicians. Although cock fighting has been banned for over a century, there is still an active trade in game birds and they are often stolen. There are two types, the Oxford which is nearer the original, and the Carlisle with broader breast and horizontal back. The Oxford has around 30 colour variations while the Carlisle has 15. There is a bantam form of the Oxford type.

Old English Pheasant Fowl (See p. 41)
An old breed, this has also been called the Yorkshire Pheasant in the past, when it was a popular farmyard bird in the region. It lays white eggs. There is a Gold variety and a Silver one, as well as a bantam version.

Orloff (see page 42)
Originally from Iran before being taken to Russia, the Orloff is a heavy breed with a small comb and considerable muffling around the face. It lays tinted eggs and is available in the following varieties: Black, White, Spangled, Cuckoo, and Mahogany. There is also a bantam form.

Cuckoo Marans.

North Holland Blue

Minorca

Pure Breeds

Gold Laced Wyandotte

Old English Game

Left to right: Silkie, Rhode Island Red, Silver Sussex and Marans.

Silver Spangled Hamburgh

Black Orpingtons

Marsh Daisy. The breed has greenish legs.

Jubilee Indian Game

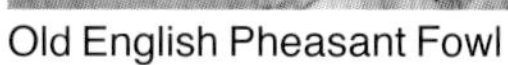

Old English Pheasant Fowl

New Hampshire Red male.

Transylvanian Naked Neck.It carries a gene for reduced abdominal fat when crossed to produce table birds.

Lakenvelder

Vorwerks with their attractive 'velvety' plumage.

Norfolk Grey

Chamois Poland

Orloff

Plymouth Rock (see page 37)
A heavy American breed, the Barred Plymouth Rock is often crossed with the male Rhode Island Red to produce a black sex-link cross for egg and table production. In addition to the Barred, there are Black, Columbian, Buff, White and bantam.

Poland (see also page 36)
Originating in Poland, this is a light breed which can be flighty if not confined. It lays white eggs. Varieties include White Crested Black, Chamois (seen left), Gold, Silver, Self-White, Self-Black, Self-Blue, White Crested Blue, and White Crested Cuckoo. The White Crested varieties, unlike the others, do not have muffling. There is a bantam.

Rhode Island Red (see opposite and page 40)
Traditionally one of the most important American, dual-purpose breeds. Most brown egg laying hybrids are based on it. It was frequently crossed with breeds such as Light Sussex and Barred Plymouth Rock to produce crosses that could be identified at hatch. Eggs are mid-brown. There is a bantam.

Rumpless Game
Bred in Britain from a genetic mutation of Old English Game, the Rumpless is so-named because of the absence of a tail. There is a bantam version which is more common than the large fowl. Eggs are tinted.

Scots Dumpy (see page 22)
An ancient breed, this has short legs and a waddling walk. Eggs are white in colour. Varieties are Cuckoo, Black, White, Brown, Gold, Silver, and bantam.

Scots Grey
Bred in Scotland, this steel-grey and black bird lays white eggs and needs extensive, well-fenced grazing for it can be flighty. There is a bantam.

Shamo (see page 46)
This is a large, upright-stanced, Malay type bird bred in Japan for fighting. It lays white eggs.

Sicilian Buttercup (see page 37)
Thought to be from Sicily, it was introduced into Britain early this century. It lays white eggs and its most distinguishing feature is the prominent, cup-shaped comb. There are two varieties, the Golden and the Silver. There is also a bantam.

Silkie (see pages 40 and 46)
A large fowl despite its bantam appearance, it originated in Asia and lays tinted to cream eggs. The plumage is very fluffy, resembling hair rather than feathers. The hens regularly go broody and are often used to hatch the eggs of other breeds, in the absence of an artificial incubator. In addition to the White variety there are Black, Blue, Gold, and Partridge varieties. There is also a Bearded Silkie with beard and ear muffling. There is a bantam.

Spanish (see pages 25 and 46)
A light, Mediterranean breed, the Spanish has glossy black plumage with a white face and lobes, while the comb and wattles are red. It lays large, white eggs.

Sulmtaler
A heavy Austrian breed, the Sulmtaler was bred from local table fowl crossed with Houdan, Dorking and Cochin. A noticeable feature is the tuft of feathers behind the comb. Eggs are light brown. There is a bantam version.

Sultan (see page 47)
Hailing from Turkey, the Sultan is a suitably exotic light breed with a striking crest, beard and feathered legs. The plumage is pure white and the eggs are also white.

Rhode Island Red

Rumpless Game

Scots Grey

Hybrid Layers bred for Free-Range

Bovans Nera

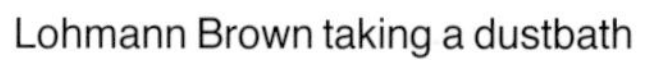

Lohmann Brown taking a dustbath

Black Rock at the back and Calder Ranger (Columbian Blacktail) at the front

Speckledy layers. The one at the front shows the first indication of the annual moult

Commercial flock of Hisex Rangers

White Star layers of white eggs.

Hybrid Table Birds bred for Free-Range

Hubbard-ISA red-feathered table birds.

Cotswold Whites

Ross and Cobbs, although bred for intensive production, will adapt to outside rearing.

One of the Sasso red-feathered strains.

Two lots of young table birds about to go outside after initial rearing inside under a heat lamp.

Shamo

Trio of the author's Silkies

Spanish *(Hicks)*

Sumatra *(Edward Brown).*

Sumatra
These mysterious looking, upright birds from Asia have a partridge-like appearance with a long, flowing tail. Feathering is greenish black. Eggs are white. They are to be found in two varieties: Black and Blue which is a slate-blue colour. There is also a bantam.

Sussex (see pages 36, 37 and front cover)
This was an important commercial breed in the past, used as a table bird and as an egg producer. The Sussex is a heavy, British breed laying tinted eggs. Varieties are Speckled, Brown, Buff, Light, White, Silver, Red and bantam. The Light Sussex was often crossed with the Rhode Island Red male to produce easily identifiable male and female chicks at hatch. The males of the cross are silver-yellow while the females are yellow-brown.

Transylvanian Naked Neck (see page 41)
Originating in Romania and introduced to Britain in the 1880s, this extraordinary breed with its neck denuded of feathers, is sometimes crossed with other breeds in the table bird industry because of a gene for reduced abdominal fat. Eggs are tinted. There is a Black, White, Buff, Red, Blue, and Cuckoo.

Vorwerk (see page 41)
A light breed from Germany, this has attractive velvety plumage of buff and black. Eggs are tinted to cream. There is a bantam version.

Welsummer
A light breed from the Netherlands, which lays dark brown eggs if the strain is a good one, otherwise they are mid-brown. It was introduced into Britain in the late 1920s. There is a Silver Duckwing variety as well as a bantam.

Wyandotte (see also pages 37 , 40 & 66)
A heavy American breed laying tinted eggs this was once an important utility breed for eggs and table. It is available in many varieties, including Barred, Black, Blue, Buff, Columbian, Partridge, White, Silver-Pencilled, Silver-Laced, Red, Gold-Laced, Blue-Laced, and Buff-Laced. There is also a bantam.

Yamato-Gunkei (see page 50)
The largest of the small Japanese game birds, this is a rare breed producing tinted eggs. In appearance it is thickset with sparse plumage. It is sometimes regarded as a bantam, but it is not a true bantam.

Yokohama
The striking characteristic of this Japanese bird is its long tail. In Japan, this is kept growing to inhumane lengths, with the unfortunate bird kept in conditions where its movements are restricted. Under normal conditions, the feathers would moult before reaching excessive lengths. There are Golden, Duckwing, Red-Saddled, and White varieties. There is also a bantam.

Sultan

Welsummer

Columbian Wyandottes

Yokohama

True Bantam Breeds

Barbu d'Uccle Porcelaine - Belgian Bearded

Black Rosecomb

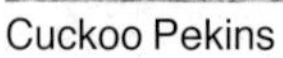

Cuckoo Pekins

Nankin

Silver-Laced Sebrights

The author's Dutch bantam male.

True Bantam Breeds

Although many large fowl breeds have small versions, also called bantams, the following are the only true bantams for which there are no large counterparts.

Belgian Bearded (see also page 54)
Belgian or Barbu bantams are available in a range of plumage types and colours, including Quail, Porcelaine, Millefleur, Blue, Silver, Lavender, Cuckoo, Black, Black Mottled, Citron, and White. (Barbu means beard). Barbu d'Anvers and Barbu de Watermael do not have feathered legs. There is a Rumpless (no tail) form of both d'Uccle and d'Anvers.

Booted
This old, feather-legged breed was used to produce the Barbu d'Uccle by crossing with the Barbu d'Anvers, the clean-legged members of the Belgian Bearded group. Most are either Black or White, but there are also Black Mottled, Millefleur and Porcelaine.

Dutch
Laying tinted eggs, these little birds are popular, if vocal members of the farmyard, as well as of the show circuit. There are many varieties, including Gold Partridge, Silver Partridge, Blue Partridge, Blue-Silver Partridge, Yellow Partridge, Blue-Yellow Partridge, Crele, Pyle, Cuckoo, Black, White, Blue, Lavender.

Japanese
Noted for their small stature, very short legs and upright tail, Japanese bantams lay white-cream eggs. There is a wide variety of colours: White, Black, Black Tailed White, Black Tailed Buff, Buff Columbian, Birchen Grey, Silver Grey, Dark Grey, Miller's Grey, Mottled, Blue,

Porcelaine Belgian Booted bantam hen.

Lavender Barbu d'Anvers

Japanese bantam

Tuzo, the only hard feather true bantam.

Yamato-Gunkei, often regarded as bantams but they are not true ones.

Old English Game bantams, a miniature form of the large fowl.

Lavender, Cuckoo, Red, Tri-Colour, Black Red, Brown Red, Blue Red, Silver Duckwing, and Golden Duckwing.

Ko-Shamo (see page 59)
This is a Japanese bantam game bird with sparse, although hard plumage, but it is not a true bantam. The male has a particularly erect stance. There is a range of colours, including Black, Black and White, Black-Red, Blue, Buff, Duckwing, Spangled, Cuckoo, and White.

Nankin (see page 48)
Nankins are some of the earliest bantam introductions from the East, and are docile yet quite good layers of tinted eggs. They are clean-legged, with yellow-orange plumage (named after Nankin cloth).

Pekin (see pages 33 and 48)
Originating in Asia, the Pekin is a very pretty short-legged, fluffy bird with feathering on the legs. It produces white or cream eggs. There are White, Black, Blue, Buff, Mottled, Barred, Columbian, Cuckoo, Lavender and Partridge varieties.

Rosecomb (see page 48)
Bred in Britain, the Rosecomb is a glossy, cobby little bird that lays white-cream eggs. There are Black, White, and Blue varieties.

Sebright (see page 48)
Another British development, the Sebright is noted for the lovely lacing pattern on its plumage. It lays white-cream eggs, but not in any great numbers, and fertility has undoubtedly declined as a result of breeding for the upright tail, a factor which is often associated with lack of productivity. There are two varieties, the Silver and the Gold.

Tuzo
From Japan, the Tuzo is a hard-feathered breed, rather like the Asil in its upright stance, broad front and small comb. Eggs are tinted .

Hybrids for Free-Range

These are commercial strains, some of which have been developed for egg laying or as table birds for free-range and organic production. Layers bred for free-range are slightly heavier than cage hybrids so that they are able to cope better with outside conditions. Table birds for free-range grow more slowly so that they are less likely to develop leg weaknesses sometimes associated with intensive broilers. They are also available in a range of different feather colouring. See pages 44 and 45 for illustrations of these breeds.

Free-Range Layers

Babcock B380: Bred for free-range, the brown feathered Babcock produces 300+ mid-brown eggs.

Black Rock: This is a black sex-link cross of the Rhode Island Red and Barred Plymouth Rock, producing 280+ eggs. Although soft-feathered, it has been bred to some extent for tighter feathering to cope with rain.

Bovans Nera: Another back sex-link of the same cross, bred for extensive conditions. Developed in the Netherlands, the birds produce 270-290 brown eggs. There is also a lighter *Bovans Goldline* laying 300+ eggs.

Calder Ranger: Originally called *Columbian Blacktail*, this has been bred for docility, as well as for free-range conditions. Feathering is brown with black tail feathers. 290-300 eggs.

Hebden Black: Another black sex-link bird of the RIR x BPR cross, the Hebden Black produces 260-270 dark brown eggs.

Hisex Ranger: This has been developed for outside conditions and is slightly heavier than the Hisex Brown which is also popular for commercial free-range. Egg numbers are 290-300.

Lohmann Brown: A docile layer of around 290-300 light brown eggs. There is also a *Lohmann Tradition*. Both are bred for free-range.

Speckledy: This is a commercial hybrid based on the Maran to produce around 290 brown, speckled eggs.

White Star: Based on the White Leghorn, this lays 300 + white eggs.

Free-Range Table Chickens

Cobb 500: Although bred for indoors, the white-feathered Cobb does adapt to free-range as long as it does not have too much protein in the diet, and has enough exercise.

Ross Table White: Also bred for indoors, the Ross can adapt in the same way as the Cobb.

Euribrid Hybro: These are white feathered birds bred for slower growth, for free-range.

Sasso: French Sasso strains are bred for slower growth and are produced by crossing pure breed males with female hybrids that have a recessive gene so that the progeny always resemble the fathers, yet are short-legged with plenty of breast meat. The breeders also produce layers developed in a similar way from traditional breeds. Strains are available under different names.

Hubbard-ISA: Like the Sasso birds, these strains are also available in different colours - white, brown, grey, black, etc, to suit particular markets. They are often given local names by the breeders who distribute them.

Buying the Birds

Chickens, she said, were worth three shillings
a couple in their own good town of Peniworth.
(The Chicken Market - British Fairy Tale)

Once the house is set up, and choice of breed decided upon, the next stage is to acquire the chickens. There are several issues to consider before buying them, however. They include *age, availability, health* and *cost.*

Age

Fertile eggs This is the cheapest way of buying, although transport costs may be fairly high. It is not recommended for those new to chickens. The eggs may be infertile, the hatching rate may be low, there may be more cockerels than females, an incubator will be needed and the chicks need protected conditions until they are hardy. (See *Breeding).*

Day old chicks Day-old chicks have remnants of yolk in the abdomen and so have enough food to be going on with for a couple of days. This means that they can be transported without too much difficulty, and without incurring the more stringent requirements of the transport legislation that affects older birds, and which subsequently pushes up the delivery cost.

The chicks will hopefully have been sexed. Even so, sexing is a difficult procedure and it may be that what you thought were all females, turn out to have some males among them.They will need a sheltered area where rats cannot get in, and where an overhead heat lamp provides the necessary warmth. (See page 74). The area can be a partitioned area of a shed or spare room, or even a large cardboard box, depending on numbers. There are also purpose-made brooders available from poultry suppliers.

The chicks will need to be confined to the area near the lamp and have a layer of poultry wood shavings as flooring. The correct height of the lamp will be established by looking at the behaviour of the chicks. If they cluster in a tight ball in the centre, they are cold and the lamp should be lowered. If they are all ranged at the periphery, they are too hot and the lamp should be raised. If they are spread out evenly, the height is about right. As they grow, the lamp is gradually raised until it is no longer needed. The chicks are normally hardy at around six weeks old.

They should be given chick crumbs from hatch because it is the right size for them. Some brands contain an anticoccidistat for the prevention of coccidiosis, a disease that can be fatal to chicks. On a small scale, if care is taken to prevent areas of damp on the flooring, and the chicks are in an area that has not been used by poultry before, they are far less likely to get it.

Growers Growers, from the age of 6 weeks onwards, will be well grown, fully-feathered and hardy. They will be more expensive than day olds but the survival rate will be higher. From the age of 6 weeks they can be weaned off chick crumbs and given a grain ration, as well as poultry grit to keep the gizzard working healthily. Special growers' rations are also available.

Point of lay pullets The easiest course of action for a beginner is to buy pullets at point-of-lay (POL). This is from around 16-18 weeks, when the young females will be coming up to the time of laying for the first time, at 19-21 weeks. It gives them time to settle down in their new home before laying begins. Most hybrids are bought at this stage. When buying hybrids, check that the birds have been perch-trained, otherwise they will not readily go to roost in their new house, and you will have to teach them. You can also request that they are not beak-trimmed. (Most commercial hybrids are beak-trimmed, but it is not necessary for birds that are kept non-intensively.) Ask the supplier what the pullets have been fed on. If they are still on a grower's ration, this can be gradually replaced with layer's feed.

Older birds If you live close to a commercial battery farm, you may be able to buy battery hybrids at the end of their period of commercial laying (one year). They will be very cheap, virtually lacking in feathers and will not know how to perch. When first put in their house they will be reluctant to come out, never having experienced the outside world, but in a short time, they will learn to do so. Once re-feathered and active, they will be different birds and will lay well for a number of seasons. Many poultry keepers have experienced the great satisfaction of 'rescuing' battery birds in this way and restoring them to humane conditions. Be aware, however, that if you intend to sell eggs that are described as 'organic', hens from batteries cannot be used under the required standards. They may also introduce disease.

When buying pure breeds, you may be offered a 'breeding trio' of older birds, or indeed of any age. Try and avoid this because they may be closely related, a situation which can lead to birth defects if you subsequently breed from them. Always get a male from a different source.

Availability

Hybrids are generally easy to buy because there are usually local agents of the breeding companies in most areas. The cheapest option is to go and collect them yourself, for small numbers may incur relatively high delivery costs.

Pure breeds are available from specialist breeders and there is sometimes a waiting list. Unless there is a breeder in the area, it may be necessary to travel quite a distance, or to go to a show where the breeder is exhibiting. There are those who specialise in delivering small numbers of birds, but it may be necessary to wait until they have a general delivery in your area before they can do so. Pure breed and hybrid suppliers catering for the small sector, are listed in poultry magazines.

Health

Anyone buying chickens is advised to buy those that have been vaccinated against Marek's disease, and Newcastle disease, and which come from a breeding flock that has been tested for the presence of *Salmonella enteriditis.* This is a form of salmonella that is passed from the parent bird to the chick in the shell, so it is important that breeding birds are free of it for they can be carriers without showing symptoms. Hybrid suppliers and larger breeders of pure breeds will have tested and vaccinated birds, but smaller breeders of pure breeds may not be able to provide these guarantees.

Cost

There is a considerable difference in price between hybrids and pure breeds, with the latter costing up to four times the price of the former, depending on the breed. Add to this the cost of delivery, and keeping pure breeds can be an expensive option. Neverthless, many people feel that they are doing their bit to help conserve the old traditional breeds by keeping them, and it must be said that pure breeds are generally prettier and with a greater range of feather colouring than most hybrids.

Introducing birds to their new home

When they are delivered, they will be stressed, particularly if the journey has been a long one, so handling them gently and talking to them quietly is important. The correct way to hold a chicken is to support its body from below, while confining its wings from above. Having the head end towards you reduces the likelihood of getting droppings down your front.

The first priority is to put the birds into their house, with a feeder and drinker, and then leave them alone. It's a good idea to keep them confined to the house for about 24 hours before opening the pop-hole to let them out into the run. This ensures that the concept of 'home' is imprinted straight away. When it is time to open the pop-hole after their confinement, leave them to come out in their own time. Some birds emerge straight away while others are more cautious. If the feeder and drinker are now placed in a covered part of the run, they are more likely to investigate. After a short time, the birds are used to their new home, and go in and out as if they had always been there. Exceptions are likely to be ex-battery hens referred to earlier. Even so, they normally pluck up courage after a day or two, and come out into the light of day.

Barbu du Watermael Quail bantam.

Feeding

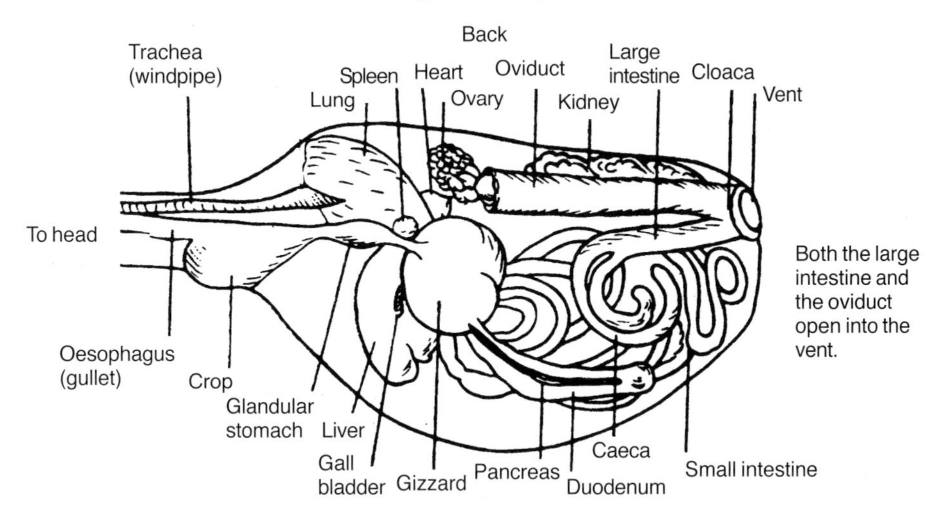

Balance is all when it comes to feeding. The digestive tract is adapted to 'little and often'. Too much of one thing may lead to a full crop, but if it is at the expense of other nutrients, the result is an unbalanced diet.

A *compound feed* is one that has been formulated to provide all the necessary nutrients, and is the choice of most poultry keepers. It is available in the form of *pellets* or as *mash* (powder form). Pellets are slightly more expensive than mash, but are more convenient. Mash needs to be stirred before being put in the feeder, in case the calcium element in it has sifted to the bottom.

The average laying hybrid will take in 130g of compound feed a day. (Larger birds and those that are free-ranging will have a higher intake).

Many people are concerned to know what goes into a compound ration. Since the BSE tragedy, no animal proteins, fats or animal by-products are allowed to be used in poultry feeds. Fish protein is allowed, but many people prefer to avoid 'farmed fish' protein because of the colouring agent that is put in fish foods in order to make the flesh pink (farmed trout and salmon). Wild sea fish protein is preferred but even so, too much can impart a fishy taste to the eggs.

Most free-range feeds contain plant proteins, the bulk of them being composed of grains such as wheat, maize, oats and barley. Soya or crushed peas and beans may also be used as these are high in protein. Where soya is an ingredient, full-fat soya rather than extracted soya is preferable because concerns have been expressed about the possible carcinogenic effect of the acids used in the extracting process.

Elements of a diet: grain, compound pellets and crushed oystreshell. Insoluble grit is also needed to keep the digestive system healthy.

An outside feeder which can be raised or lowered to suit the size of birds and can be used for on-demand feeding of grain or pellets without the risk of wastage by wild birds or rodents. *(Hengrave Feeders)*

Many compound feeds contain artificial additives; some necessary, some not. A mineral and vitamin supplement is necessary, with the trace elements being a tiny but essential part of the ration. Yolk enhancers, such as *Xanthophyll,* on the other hand, are definitely not needed. These are added in order to make the yolk more yellow. (Caged birds with no access to grass and the sunshine would have very pale yolks if they were not artificially coloured via the feed). Broiler feeds for table birds often contain low levels of antibiotics, a practice which has been condemned by medical authorities because it encourages the growth of antibiotic resistant diseases. Avoid them!

Genetically modified soya and maize have also been introduced in recent times, causing concern to many. The modified soya for example, has a resistance to herbicides leading to worries that if it 'escaped' into the wild, new strains of 'super weeds' might develop. Modified maize has a resistance to the antibiotic *Ampicillin,* raising the spectre of antibiotic tolerance which encourages new strains of antibiotic resistant pathogens. Some feeds are described as 'natural' because they do not contain antibiotic growth promoters or yolk enhancers, but they may still have extracted soya and genetically modified grains. The term 'natural' has no legal definition. Organic feeds *are* legally defined, and contain no artificially extracted soya or genetically modified ingredients. They are more expensive than other feeds.

Free-ranging hens add natural minerals to their diet. This one is a Speckledy.

In some areas they may be difficult to find, but they are becoming more common, and are the best rations available for poultry. If they are not available in your area, buy a compound ration that is described as a 'natural, free-range feed', and try to obtain an assurance from the supplier that no unnecessary additives have been included.

Grain

Although the compound feed does provide all the elements of a complete ration, it is a good idea to feed a separate grain ration such as whole wheat. This is cheaper than the compound feed, and is also appreciated by the birds. A good plan is to feed the pellets or mash in the morning and to give the grain in the afternoon. Around 15g of grain is the quantity taken by the average hybrid layer, but bear in mind that larger and outdoor birds will take more. In winter, as more demands are made on the metabolism (keeping warm as well as laying eggs), it is necessary to increase the rations. Some extra grain, is the most cost-effective.

Grit and calcium

When a hen takes in food, it is initially stored in the *crop* or holding area of the digestive system, before it goes into the *gizzard* for digestion. The crop is in the breast area and if you pick up a hen that has just eaten grain, you can feel the grains in the crop. From here food passes to the *glandular stomach* where secretions moisten and begin to break it down. It passes to the gizzard which has strong muscular walls which contract to grind up the grains. In order for this to be accomplished, there must be something to grind them against. A chicken, like most birds, will take in small pieces of grit or stones for this. If grain is being given as an extra, it is necessary to provide grit unless the chickens are free-ranging sufficiently to find their own sources.

Calcium is needed for strong shells, as well as for bone structure, and a deficiency may be indicated by fragile shells. Compound feeds contain calcium and a certain level of grit. Poultry grit and crushed oystershell are available from feed suppliers. Traditionally, egg shells were heated in the oven, crushed and fed back to the hens. There is nothing wrong with this, as long as eggs are not being sold, they are heated enough to destroy any pathogens, and that shells are crushed so as not to be identifiable to the hens. (You don't want to encourage egg eating!)

People often ask whether table scraps should be given to domestic chickens. Scraps should not be given where eggs are being sold, but there is nothing wrong with giving them to household chickens, as long as they are fresh, have no meat scraps and are not salty. Brassica plants and other kitchen garden gleanings are also popular. As referred to earlier, hanging up greens in a run does provide interest for the birds, and helps to prevent anti-social behaviour such as egg eating and feather pecking.

A Range of Feeders

Metal open feeder. This is very stable but the birds can scratch out the food which is also easily soiled by droppings.

This metal feeder has a bar across the top making it more difficult to waste food. *(Eltex)*

Plastic feeder with inserts to prevent wastage. It can be placed on the ground or suspended.

Self-dispenser that operates by displacement of a lever by the chickens. *(Parkland)*

Galvanised metal feeder suspended in a barn.

Self-dispensing feeder *(Solway)*

Some of the garden plants and weeds that chickens seem to like in particular include: Parsley, *Petroselinum crispum,* Chickweed, *Stellaria media,* Fathen, *Chenopodium album,* and Good King Henry, *Chenpodium bonus-henricus.* Do make sure that these are tied up in bundles and hung up, rather than just being thrown into the run. It is important that the birds are able to peck off small pieces, rather than having to cope with long strings which may cause a blockage in the crop.

Water

Clean, fresh water needs to be available all the time. A drinker that is raised off the ground will stand more chance of remaining clean. Droppings and scratched earth and grass can soon find their way in. A drinker can also be suspended. For larger numbers of birds, an automatic supply is not difficult to set up. Most poultry equipment suppliers have a good range of drinkers and feeders, manually filled and automatic. As a general rule, five hybrid hens will consume one litre of water a day under normal conditions. Consumption will naturally be more in hot weather, and where larger birds are concerned.

Winter can be a trying time when low temperatures freeze the water supply. The best way of preventing this is to use electrical heating tape which is obtainable from specialist suppliers. Alternatively, place the drinker in an insulated cover, and put the whole thing in a protected area. (See page 64 for more details).

Ancona bantams

Ko-Shamo bantam

A Range of Drinkers

Suspended plastic drinker. The hen is a Lohmann Brown.

Galvanised metal drinker placed on bricks to stop scratched soil getting into the water.

Metal drinker suspended in a barn. The chickens are Black Orpingtons.

Suspended plastic drinker with water from an overhead tank.

Large outdoor drinker on wooden slats. The hens are ISA-Browns.

"Those hens think this drinker's just for them!"

Daily and Seasonal Care

To everything there is a season
(The Bible)

Daily care

The first task of the day is to open the pop-hole and let the chickens out.It helps to be able to do this from outside the run. This is a good time to check for any potential problems. A hunched-up bird is a sure sign of illness or bullying, or both. (See later for advice on what to do).

Check the feeder and drinker and if necessary, clean them. Give the birds some layer's pellets (or layer's mash) in the feeder and fill the drinker. Top up the grit container with poultry grit and crushed oystershell. Check for eggs and remove them. If any of the wood shavings in the nest box have droppings adhering, remove them and replace with fresh shavings. Hang up some greens if the birds are confined to a run, to give them added interest. If necessary, move the house and run to a fresh area of grass. If the birds are free-ranging, check the fencing and if necessary, make a new area of grass available to them.

Check the birds during the day, and talk to them. They are more responsive than people think, and become quite tame towards those who feed and care for them. In the afternoon, either give some wheat on the ground, so that they have the opportunity to scratch for the grains, or make it available in a feeder. Check the water and collect any eggs. Put these in a cool place, a larder or refrigerator, as soon as possible. Keep a general eye and ear open for any unusual sights or sounds that may indicate bullying, going off to lay in the hedge, the presence of danger, and so on.

At dusk, check that they have all gone into the house and then make sure that the pop-hole and door are closed securely.

Weekly or monthly routines

Regular cleaning is important, and depending on the type of house and number of birds, this may be anything from a weekly or monthly task to one that is more seasonal. Droppings need to be removed from the house on a regular basis. If the house has a droppings board, it is a matter of sliding this out and disposing of the droppings in the compost heap. Don't let the chickens have access to the heap because that would pose a strong possibility of infection. If necessary, put the droppings in a thick plastic bag, tie it up and leave the droppings to rot down for a few months. (When it's time to do the winter bed preparation in the garden, use it as a soil conditioner and natural fertiliser). Some solid-floored houses may have a thick layer of plastic with sawdust or wood shavings on top. The whole thing can then be removed quite easily. Put clean wood shavings in the nest boxes. Remove the perches and give the whole house a good sweep out.

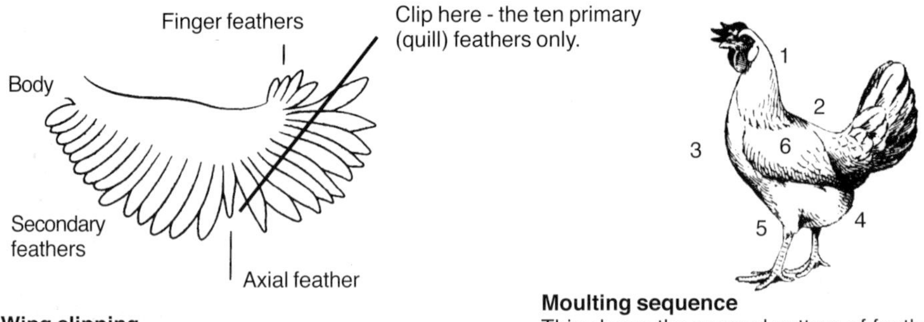

Wing clipping
As a last resort for flighty birds, the primary feathers on one wing only can be clipped as a temporary measure.

Moulting sequence
This shows the general pattern of feather loss but it can vary considerably. Hybrids tend to moult more heavily than pure breeds.

Seasonal care

Spring

This is the best time of year to start keeping chickens. They start to come into lay and the warm days of summer are still to come. It's also a good time to buy an incubator if breeding is to take place, or if you already have one, to overhaul it and get it ready for the new season. Only incubate eggs from the best birds, for there is little point in just breeding indiscriminately. That leads to poor, possibly sickly birds, and they will eat just as much as good ones.

Check the flock regularly to make sure that they are not affected by lice or mite attack, and also check the house in case red mites have taken up residence. Give the hens a daily 'look-over' in case there are any signs of feather or vent pecking, or any other indications of aggressive behaviour.

Keep an eye open for broody hens! Broodiness is the natural inclination for a hen to sit on and incubate a clutch of eggs. It is easy to spot; the hen will be hogging a nest box and will fluff up her feathers and give a peevish squawk if disturbed. She may even aim a peck at you, although this will be a half-hearted attempt. She is focused on serious sitting, the breast area will be hot to the touch and she may have plucked out some breast feathers to line her nest. If you happen to have some fertile eggs, they can be slipped under her for hatching. If she is stopping other birds from using the nest box for egg laying, she can either be moved carefully to a broody box, or if you want to stop the broodiness, placed in a temporary run in cooler conditions which will make her change her mind about the whole business.

Summer

It's a lovely time of year, but unfortunately external parasites such as lice and mites think so, too. Watch out for them, and at the first sign of trouble, take action, as detailed in the *Problems* section. The warm days also encourage the hens to go further afield, and if they are not confined to a run, may decide to lay in a 'wild' place such as under a hedge. They may also be attracted to the idea of perching on low tree branches. Some of the old, light

Check the hens during the day and talk to them. They are more responsive than people realise and will soon acknowledge the one who feeds them as 'head of the flock'. These are Leghorns.

breeds have a greater tendency to 'fly' than other birds. If you have birds that are great escapists, it may be necessary to trim the flight feathers on one wing. This causes loss of balance when they try to get over a fence, but causes no harm, although they are less able to escape if danger threatens, so must be protected.

Trimming the flight or primary feathers is not difficult, and sharp shears or scissors can be used. It is quite painless (equivalent to trimming our nails) and feathers do grow again, but it is unsightly and not recommended for those who keep show breeds. It is important to trim only the ten primary feathers, not the secondary ones, otherwise the bird's ability to keep warm is hampered.

Mid to late summer is the time of year when moulting occurs. This is a natural loss of feathers, to be replaced by new ones. Birds will normally go though the first year without much moulting, but in the second year it is usually more noticeable. It can vary quite dramatically between different birds and breeds, with hybrids often displaying quite a dramatic moult, while pure breeds may just lose a few feathers at a time. Egg laying may also stop at this time. Ensure that hens have a balanced diet (feathers are mainly protein) and a lack may delay re-feathering. If moulting is prolonged, lasting for many weeks, there may be some other reason such as stress or an attack of mites. (See *Problems*).

Stopping the water from freezing

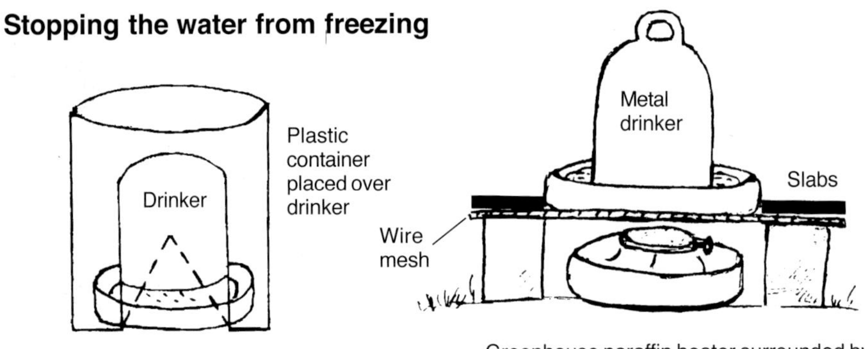

Notch cut in front for access

Where the frost is not too extreme, providing an outer container as protection may be enough.

Greenhouse paraffin heater surrounded by bricks to prevent poultry access

Autumn

As the days draw in, the number of eggs also drops, for the duration of light affects the egg-producing system of the hen. Older breeds may stop laying altogether, and it was common in the past for ducks to be used as winter egg providers, to make up for the shortfall. Modern hybrids have been selected and bred for more consistent production. They are more reliable winter layers but will still tend to lay less unless their day is artificially lengthened by the provision of artificial light.

Many poultry keepers will not be too bothered about a decline in the number of eggs during the winter, as long as there are a few intrepid souls who continue laying. Those who are selling eggs, however, will want to ensure that there is not a marked reduction in the egg output. The best way of doing this is to have a new flock coming to point of lay as the first flock is moulting. These birds will then lay though the winter, giving time for the original flock to re-feather and recuperate.

If birds have ceased laying as the days are getting shorter, it is customary for commercial producers to give them some artificial light to make up the deficit. This is provided before dawn, after dusk or a combination of both, to suit the producer. A system which can be automatically adjusted with a time switch is normally used, but providing lighting in a small moveable house is more difficult than doing so for a larger shed where mains electricity may be available. Fortunately, there are now quite sophisticated, electronic systems for the small poultry keeper. These measure the day length and light comes on automatically in order to keep a standard day length. No manual adjustments are required once the electronic sensor has been placed where it can 'see and learn' about the availability of natural daylight.

Placing a moveable house and run near the house is sometimes enough to provide winter eggs. Just delay drawing the curtains for a while, so that the light shining out is enough to make up the deficit.

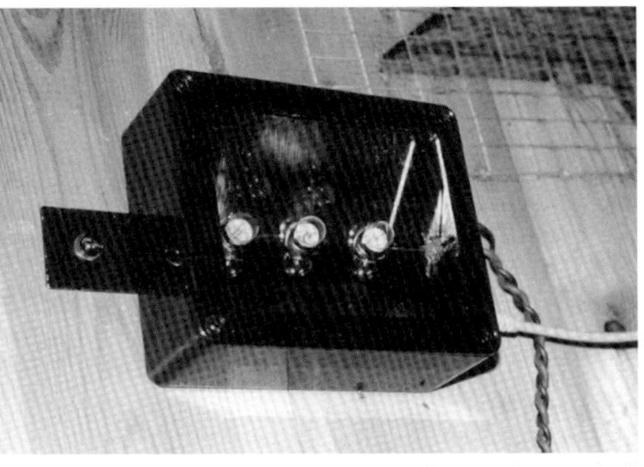

Rooster Booster bolted to the wall near the roof of a small house. The power lead goes down to a car battery to which it is attached by two clips. The sensor lead going off to the right passes through a ventilation hole and is installed under the roof eaves where it can 'see'.

Artificial light for small houses
Above is the *Chrysalis* fluorescent light mounted in a poultry house and connected to a 12v car battery. The sensor on the right is temporarily taped to the window glass before being placed outside, under the eaves.

Right: Light sensor under the eaves adjusts automatically to availability of natural light after 24 hours 'learning'.

Should one worm hens at this time? This is a question that is often asked. If they are moved regularly to fresh ground and over-stocking is avoided, the answer is usually, no! Internal, parasitic worms can be a problem however, particularly for free-ranging birds in an area where pheasants are common, but the signs are easy to spot. A thin hen with a razor sharp breast bone, excessively bedraggled plumage (not to be confused with the natural moult), anaemic looking comb and wattles and a permanently mucky bottom, is likely to have worms. Pick her up and see if she feels lighter than usual. If you are used to handling your hens, you should be able to detect changes in weight quite easily. Details of the worming are in *Problems.*

Winter

Birds with particularly large combs and wattles are at risk of frostbite. In order to avoid problems, make sure that the house is reasonably well insulated and that the birds are not let out too early. Prominent combs and wattles can also have a little *Vaseline* rubbed in to protect them. It works well, but if a bird has already been caught by frost and the comb is badly damaged, it may never recover. Severe frost-bite damage needs to be trimmed back, but a vet should do this.

Every effort should be made to keep floor litter clean and dry so that there is less likelihood of diseases such as coccidiosis. These are Barred Wyandotte bantams.

The optimum temperature for a chicken is 21°C, which is quite high - akin to a warm summer's day, and a reminder of the chicken's jungle origins. When winter temperatures plummet, extra feed is needed to cater for the body's basic metabolic needs, as well as for extra demands such as winter egg production. Extra layer's pellets can be given, but this is an expensive option, and it is cheaper to give some extra grain such as wheat. If the birds will take some oats as part of the grain allocation, so much the better, for it is the most warming of the grains, as those of us who like our winter porridge will testify. Some birds are better than others at taking oats. Those reluctant to try anything new may need to have some kibbled (chopped rather than ground) oats hidden amongst the other grain.

Stopping the drinking water from freezing can be a problem, and it may be necessary to refill the drinker several times a day. Don't do what one thoughtless poultry keeper did, and add toxic anti-freeze to it! He killed all his birds. Some suggestions for DIY solutions are shown on page 64. Alternatively, use soil heating cables that are widely available in garden centres. Be sure to use a safety circuit breaker if mains electricity and a transformer are used. It goes without saying that heat sources should be inaccessible to the chickens.

Rats and mice can be a problem at this time of year, if food is left lying around, although a moveable house that is moved regularly is unlikely to attract rats. It is static ones that are the targets, particularly if there is a build-up of droppings that send out their characteristic smell. If there is a rat problem, clear away any piles of rubbish and cut down long growth and brambles which might be obscuring their runs (they always tend to follow the same routes).Traps are available from a number of suppliers. Alternatively, rat poison such as *Sorex,* can be placed in such a way that the birds, pets and of course, children, cannot gain access to it. The local authority will come and do this for you, free of charge, if you are an ordinary householder. If you are a registered smallholder, however, you will need to pay or do it yourself.

Mice are not normally a threat to the chickens themselves. Indeed, if an unwary mouse does venture into the hen-house or run it is likely to be despatched fairly quickly by the birds. It is when they manage to gain access to feed stores and out-houses that they are a nuisance. Again, poisons and traps are available from poultry suppliers, although a cat that is a good mouser is a boon. A Siamese or part-Siamese female is one of the best mousers available. In this context, it is worth mentioning the experience of a lady who, although she had placed poison out of reach, one of her cats was poisoned by catching and eating a mouse that had taken poison. It's difficult to cope with a situation like this. The best approach is to try and avoid a mouse problem in the first place - perhaps leaving cats to do the job themselves, or relying upon the use of strategically placed traps. Feeds should be in containers with a tight-fitting lid. Household bins are effective.

Rats and mice are usually only a problem if food is left lying around and houses are in the same place all the time. *(Sorex).*

Sometimes a small flock can be housed in alternative housing for the winter so that the birds are more sheltered during the really cold weather. A garden shed or other outhouse can quite easily be adapted for winter use, giving the opportunity to carry out repairs and renovations to their normal house. Check that the roofing is sound and that no damp is getting in. Remove the perches and nest boxes, and give them and the inside of the house a thorough clean. If there is damp coming in from the outside, wait until the house is quite dry, then paint the *outside* with an exterior water repellent. Don't use it on the interior walls or you may trap moisture inside the wood!

A moveable house for winter occupation should be placed where it has wind shelter from trees or other windbreak such a fence, hedge or nylon netting. A porch outside the pop-hole is also effective.

The point was made earlier that chickens must have regular access to fresh ground otherwise they are at risk of disease and internal parasites. A small house and run can be moved regularly, ideally every other day, but the frequency will obviously depend on the size of the house, the flock and the ranging area. If the birds are free-ranging widely and flock density is low in comparison with the ground available, it may be appropriate for them to stay there for a long period with no problems.

Once grass has been vacated by the birds, it should be well raked to disperse droppings and scarify compacted areas. A sprinkling of lime will help to reduce the acidity of the soil, while grass seed can be applied to patch up any wear and tear.

What to do with all those eggs

Mine honest friend will you take eggs for money?
(A Winter's Tale. William Shakespeare).

Spring and early summer usually bring a considerable harvest of eggs, particularly if the chosen breed has a good record of production. It is a great pleasure to go out and collect the eggs from your own chickens and use them while they are still warm. No shop egg can compete with that degree of freshness!

The eggs are best stored in the refrigerator or a cool pantry, for if they do need to last, they will do so much better at lower temperatures. Any that have been soiled can be wiped clean and the 'fridge temperature will stop any pathogens developing. It is not a good idea to store eggs in one of those rustic-type egg holders that one sees in 'rural lifestyle' kitchens, where the eggs are probably shop-bought anyway. They go off more quickly at room temperatures and if the eggs have been contaminated by bacteria, may present a health hazard. Hard boiling, does of course, render them safe and is always advisable for young children, old people, invalids and pregnant women.

Selling eggs

Sometimes there are more eggs than a family can use and a surplus soon builds up. Can these eggs be sold? They certainly can, but it is important to be aware of the following conditions:

- It is not necessary to register as an egg producer if the eggs are sold to friends and neighbours, at the door or on a market stall. They may not be supplied to a shop for re-sale unless the producer is registered.
- The eggs must not be graded and sold as *Small, Medium, Large,* etc. Again, only those who are registered may do so.
- Eggs must be laid in clean conditions, collected regularly, stored in cool conditions and sold as quickly as possible.

Remember that the hens should be healthy and not allowed to run with a cockerel in case there are fertile eggs amongst those offered for sale. Ensure that the layers have come from a salmonella-tested flock. (If they are hybrids, they will be). The birds should be fed a wholesome diet and kept in a clean environment where there is less risk of bacterial contamination and disease. Eggs that are dirty should not be washed and offered for sale; they should be kept for home use and hard boiled before use.

Preserving eggs

Before the development of highly productive hybrids and the widespread availability of refrigeration, it was common for eggs to be preserved during the glut period, in order to provide for the lean time. Waterglass was the chemical used (not to be confused with *isinglass* which is used for clarifying wine in the process of home winemaking!) Some chemists still sell waterglass, but personally, I prefer more modern methods of preservation. If you are set on using waterglass, add 1 part to 9 parts of previously boiled and cooled water in a high-density, food grade plastic bin with lid, or other suitable container. Place fresh, clean eggs in the liquid as they are available, and make sure that there are no hairline cracks in any of them.

I prefer to freeze my surplus eggs! Not the eggs in the shell, of course, or they would burst. I break the eggs and separate the whites from the yolks and freeze them. If you fill ice cube trays with whites and yolks separately, they can then be removed and put in plastic bags or containers. A little salt can be added to yolks that are destined for savoury dishes, and a sprinkling of sugar for those that are to go in sweet things. Sugar and salt stop the yolks becoming sticky when defrosted. Don't forget to label them appropriately! When they are removed for use, two white squares and one yellow one equal one egg, but don't expect great omelettes or souffles!

Surplus eggs can be made into sponge cakes and trifle bases which are then frozen until needed, or eggs can be pickled. Cider or wine vinegar is nicer than the usual malt vinegar which is too strong, in my view. Heat the vinegar with a sachet of pickling spices and remove the latter when the vinegar has cooled. Hard boil and shell the eggs (if too fresh they are difficult to shell) and immerse in the vinegar when it is cold. They will keep for a few months.

Gum arabic can be painted on so that the shell is covered entirely with glaze. Dilute the gum arabic with the same amount of water and paint it on with a paint brush.

It is useful to be able to open and close the pop-hole from outside the run.

Breeding your own replacements

Why should one quarrel with good breeding?
(Pushkin, 1823)

Breeding is obviously not going to take place unless there is a cock available, and those with close neighbours are not advised to keep one. An alternative is to take your chosen breeding hen to visit a friend with a suitable male and leave her with him for a day or two. Bear in mind however, that some of the resulting chicks will be males. These will need to be sold, given away or disposed of, and it can be a problem to do any of these things.

Introducing females to a male

If breeding is desirable, it is obviously important to have birds that are in excellent health and are good examples of their type. A cock should be kept in his own roomy house and exercise pen to which *certain hens only* are introduced, if they are to produce fertile eggs. Alternatively, a small breeding flock or trio can be housed together permanently, and kept separate from the rest of the hens. In this way the latter can provide eating eggs without the risk of fertile eggs getting into the kitchen.

If a small breeding flock or trio is to be set up, it is a good idea to ensure that the male comes from a different line, in order to avoid in-breeding and the possibility of congenital defects. The male should be fully grown before the females are introduced, otherwise they may peck him so that he is unwilling or unable to mate with them. Introducing *them* to his house and run is good psychology for the same reason. The spurs of the male should be kept trimmed otherwise they may rip the sides of the hen. There is no truth in the old belief that spur-trimming causes infertility. It is a hangover from the old days of cock fighting when spurs were obviously prized for their length and sharpness. Soften the spurs with oil for several days before trimming and cut the ends only. By repeating this process over a period of time, you can get them down to a reasonable and less damaging size.

When hens start laying, the male will usually mate with them, and eggs will begin to be fertile from around three days after introduction. When mating takes place, a sperm must travel up to the infundibulum before it can fertilise the egg. (See the diagram on the next page). Sperms can stay alive for up to three weeks, so if the hens have been running in a mixed flock with other males, allow at least this time before introducing them to the required male, otherwise you may get unwanted breeding. Once the required mating has taken place, it is a good idea to wait for a further 7-10 days before taking eggs for incubation, so that maximum fertility is assured.

Hen's reproductive system

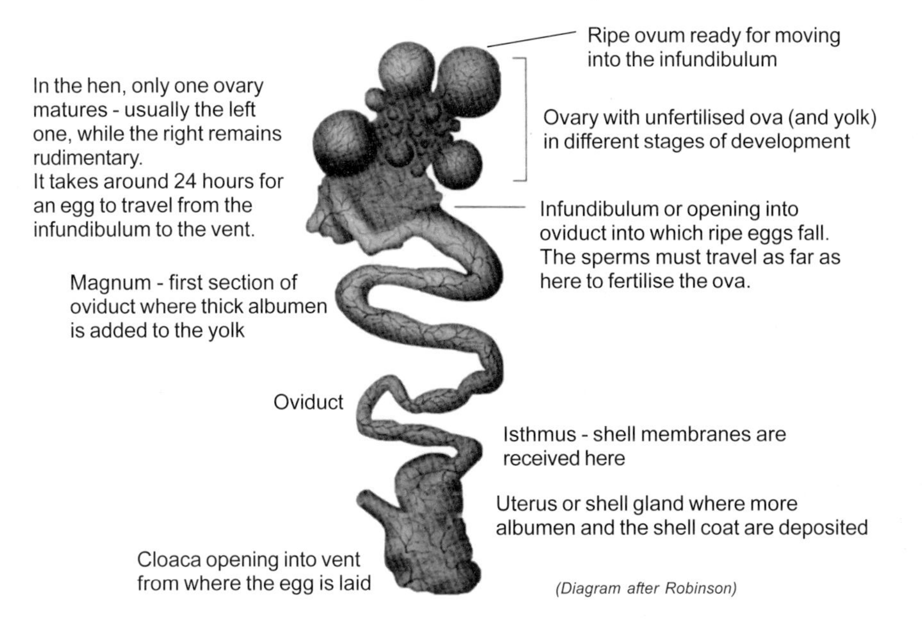

(Diagram after Robinson)

Incubation

The whole process of incubation and rearing, both artificial and natural, is covered in my book *Incubation: A Guide to Hatching & Rearing,* and this is recommended to those who are going in for breeding. Salient points are:

• Eggs should be clean and ideally not be more than a week old. After this there is a gradual decline in hatchability. If they do need to be stored for a few days, store at a temperature range of 15-18°C and a relative humidity of 75%.

• Before you incubate the eggs, wash them in warm water to which an egg sanitant has been added. The temperature needs to be higher than that of the egg otherwise dirt and pathogens on the surface may be drawn *in* through the shell pores. (Comfortably warm to the hand is about right).This will ensure that they are free of surface pathogens which may otherwise cause Mycoplasma diseases in the warm, humid conditions of the incubator. Egg sanitants are available from incubator suppliers, and their use results in more healthy chicks hatching, rather than having many dying in the shell, or soon afterwards.

• Unless you are using a broody hen to hatch the eggs, buy the best incubator you can afford. Check that it has a good insulation, adequate ventilation, an electronic temperature control thermostat, humidity meter and an automatic or semi-automatic turning facility.

Eggs need regular turning.

Chicks emerging at day 21. Avoid 'helping'.

• Make sure the incubator has been thoroughly cleaned and disinfected before use, and that it is running for at least 24 hours before introducing the eggs. This gives time to check that the thermostat is working properly.

• Place the incubator in a spare bedroom where the temperature and humidity do not vary, rather than in a shed where there can be fluctuations.

• For the first 18 days of incubation the temperature needs to be 37.5°C at the centre of the egg. After this, until hatching takes place around day 21, the optimum temperature is reduced to 37°C.

• A humidity level of 52% is needed from day 1 to day 18. This is increased to 75% from then to the hatching period.

• The eggs need to be turned several times every day until day 18. An incubator with automatic turning facilities will do this for you. A semi-automatic version means that by pulling a lever you can turn the whole batch at a time. If the incubator has no turning facility at all, it means that you will need to mark the eggs with a cross, and turn each one by hand. (Wash your hands first!)

• Follow the instructions that came with the incubator, eg, a still air machine that does not have a fan for ventilation, may need opening and closing of vents, as well as a slightly higher temperature.

Using an incubator

Incubators are either still air or forced air (draught). The former is where ventilation relies on the natural circulation of air within the box. There should be sufficient air holes which can be opened or closed as needed to control the flow. The latter is where a fan is provided to drive the air round. With this, the temperature is the same throughout the incubator. With a still air model, you need to check the temperature in different parts of the box, otherwise there may be cold spots. Follow the instructions that came with the particular model.

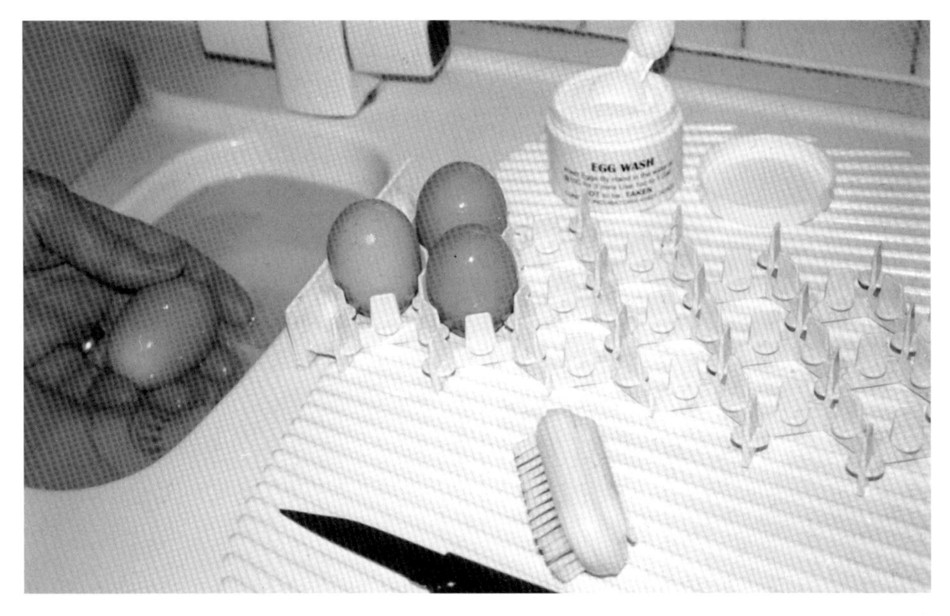

Washing eggs with an egg sanitant before incubation helps to prevent Mycoplasma infections.

The process of incubation takes around 21 days but after about a week it may be appropriate to 'candle' the eggs. This is the process of holding them up against a bright light so that the internal contents can be viewed. The original candlers were candle-powered, hence the name, but now electric-powered candlers are the norm. Make your own candler by putting a torch in a cardboard box with a hole in it, and placing the egg on the hole.

If the egg is developing normally the contents will resemble a starfish with a radiating pattern of blood vessels. Candling enables infertile eggs to be discarded so that they do not 'go off' in the incubator.

Candling also indicates whether development is proceeding normally, and that the humidity level is correct. The illustration indicates what the correct size of the air cell in the egg should be at the appropriate number of days. If it is too small, the humidity is too high and not enough moisture is escaping from the egg to allow for adequate development. If too large, the humidity is too low.

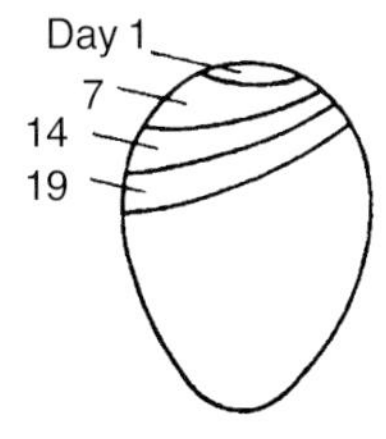

Left: The air cell in the egg should be at the size indicated for the appropriate day in the incubation period. If it is too small, humidity is too high. If too large, humidity is too low.

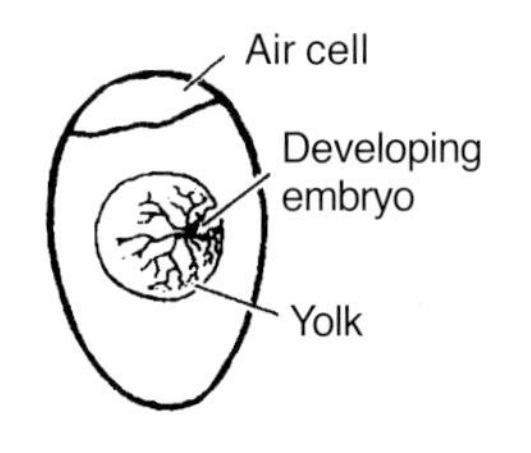

Contents of fertile eggs when viewed against a bright light. Any eggs in the incubator that show no sign of development (clears) can be discarded.

The chicks have been moved to a brooding area with a heat lamp to keep them warm. They have chick crumbs in a shallow feeder at the back. Note how the drinkers have been placed on wire mesh so that any spilt water goes through, preventing the wood shavings litter to which the chicks have access, becoming wet. This helps to prevent diseases such as coccidiosis.

Turning takes place only until day 19 of incubation because at this time the embryo will be taking up its position for breaking out. First it makes a small hole in the membranes and shell, a process known as 'pipping', and then breaks out. Average hatching is at 21 days, but may occur earlier or later.

Avoid 'helping' chicks out of the shell and leave them be until they are actively interested in what is going on around them. At this stage, they can be removed and placed in a brooder, as shown above.

A drinker will be required for water, and a shallow feeder to hold chick crumbs. If they seem reluctant to feed, place a sheet of paper on the wood shavings and sprinkle some chick crumbs on it. This is an emulation of what the mother hen does to encourage her chicks to eat, and they will come running to investigate. Reluctant drinkers can also be gently picked up and their beak tips dipped in the water so that they know where and what it is. Take care that the water does not go into their nostrils.

The broody hen

If a reliable broody hen is available, she can do the job of sitting on and hatching the eggs, as well as brooding the chicks. Broodiness can be encouraged by placing some pot (artificial) eggs in a nest box with wood shavings, in a small house. Breeds with a greater tendency to broodiness, such as Silkies and Pekins, are more likely to be encouraged in this way, than hybrids whose tendency to broodiness has to a certain extent been reduced. Once a hen is broody and sitting tight, fertile eggs can be slipped under her, while the artificial ones are removed. Do this gradually, moving one at a time, so that there is no sudden influx of cold eggs. After some shuffling she will usually accept them, but it is worth keeping an eye on her in case she decides to leave them.

"Go away! I think I'm broody".

A broody hen needs to be treated against external parasites, for she is more at risk than usual in the sedentary period. Drinking water should be placed nearby. She will probably get off the nest once a day to feed, exercise and pass droppings, and it is important to feed her close to the nest. It goes without saying that she should be in a protected area where she is safe from foxes and rodents. Any of the small broody houses with attached run are suitable. (See Pages 4 and 19). Once the eggs have hatched, the broody will look after them, picking up food and clucking to them to come and pick it up. The chicks soon learn from her, including the recognition of a warning note which has them running to hide beneath her. A mother hen and chicks is one of the nicest sights in Nature.

Earlier in the book, it was mentioned that chick crumbs provide the best start for chicks. Some brands have an additive for the prevention of coccidiosis, but it is possible to buy ones without this. Ensure that the house and pen are moved regularly so that droppings do not build up, and the ground is fresh and dry. Avoid putting it where chickens have been ranging.

It is possible to buy chick feeders with a small access so that chicks can feed from it but the hen cannot get at the crumbs: she will have her own feed. On a small scale, it may not be easy to stop them eating each other's rations, and it's not crucial. Grain can be introduced gradually, and certainly by the time they are 6 weeks. At first, it may need to be chopped or flaked.

Sexing chicks

Sexing at day old can be difficult unless the feather colouring and patterning are distinctive. Trying to examine the vent is not recommended to anyone other than an expert for it is easy to damage the young birds. Here are some guidelines for distinguishing by down feather colouring:

Any gold male x Any Silver female (except White Leghorn which is not true silver) eg, Rhode Island Red x Light Sussex = Silver male chicks and gold-buff females. (LS x RIR does not work; it produces all silvery-white chicks).

Any Black male x Any Barred female (eg, Australorp x Barred Plymouth Rock) = Male chicks all have a head spot.

Any Gold male x Any Barred female (eg, New Hampshire Red x Maran) = all chicks have head spot, but that of the male extends down the back.

Showing

Advise if this be worth attempting, or to sit in darkness here hatching vain empires.
(Paradise Lost, Book 11, Milton, 1667).

Showing chickens is an activity that many poultry keepers enjoy. It is a chance to display particularly good birds, as well as to make contacts with other breeders and potential customers, if progeny is to be sold. There are certain requirements, of course. The birds must be pure breeds that are recognised by the *Poultry Club of Great Britain.* Each breed has its own breed club which is affiliated to the national society. (Breeds that are particularly rare may not have their own club, but they come under the auspices of the *Rare Poultry Society).*

Each breed club has drawn up a *Standard of Excellence* for its particular breed, thus providing a 'template' of the ideal bird. Each aspect of the breed is considered, with a certain number of points being allocated for each characteristic such as type, colour, plumage pattern, stance, head, legs and feet, and so on. 20 points might be allocated for colour, while 5 points are given to the head, etc, with the total amounting to 100. When a bird is judged, it is awarded a number of points for each category, and the higher the overall percentage, the better the bird. Needless to say, any bird that is to be entered for a show must be in good health and condition, with nothing added or taken away (eg, clipped feathers), although giving the bird a bath and helping it to make the best of itself is allowed.

Obtain a copy of the *Standards* for the particular breed from the secretary of the particular breed club. *Standards* for all the recognised breeds are also published in the book *British Poultry Standards.* If you have a chicken or several birds that are good examples of the breed, find out which shows are being staged, when and where. The Show Secretary will have produced a *Show Schedule* which has details of the various *Classes* where poultry may be entered. These will include several

Silver Sussex bantam cock at the *National Poultry Club of Great Britain Show, 1998.* Each show cage has litter, a drinker and a notice indicating the breed, class in which it is being entered, and the owner's number.

Bath time before the show. Note the position ot the fingers to ensure that the wings cannot flap.

categories such as *Light Sussex Pullet, Bantam Hard Feather, Egg Exhibit, Large Fowl Female, Trio,* and so forth. There are also *Juvenile* classes for children. There is a date by which time entries must be received by the secretary. Once the schedule has been received, the entry forms need to be filled in and returned with the appropriate fee.

Identification of a show bird is obviously important, for it is not unknown for a bird to be put back in the wrong cage at a show, and unfortunately, theft is not unknown, particularly where game fowl are concerned. The *Poultry Club* operates a leg ringing scheme with rings that cannot be removed from the bird's leg, and where the details are registered with them. If a ringed bird is subsequently sold, the change of ownership can then be notified. Plastic rings which can be removed, are also available in different colours from poultry suppliers. These are used to identify breeding stock, particular flocks, and so on.

It is a good idea to accustom birds to being handled, as well as being transported in cages or show boxes. A bird that panics when anyone comes near its cage at the show is not going to show itself to the best advantage, and the judge is unlikely to be impressed. If stance is a particularly important aspect for the breed, get it used to standing for you. A little training goes a long way and a bird that is stroked gently along its back with a short piece of bamboo will soon adopt the necessary pose.

"I'm fed up. I want to go home!"

"Sour grapes. We won!"

A bath before the show is a good idea, particularly for light coloured breeds. A little baby shampoo in warm water works well, with a second bowl of warm water for rinsing. The illustration on the previous page shows how the wings are confined, for there is nothing worse than a flapping bird throwing caution and water to the winds. The eyes and face can be cleaned just with warm water and a small piece of sponge. The legs can be given a gentle scrub with a nail brush or old tooth brush, while the claws should be inspected for any impacted dirt. Pat the bird dry with a soft towel and, if necessary, dry it with a hairdryer on the lowest setting. This should only be used with soft feather breeds; hard feathered breeds are best left to dry naturally. Make sure that the current of air goes with the feathers.

After a bath, a little baby oil can be rubbed into the comb, wattles and legs, taking care not to get any on the feathers. Dark feathered breeds may not require bathing, but the plumage can be enhanced by using a piece of silk to rub along the natural line of the feathers, bringing up a glossy sheen. White birds should always be provided with shade if they are normally outside, for bright sunlight can have a yellowing effect on the plumage. A good, well-balanced diet is essential for tip-top condition.

Poultry need to be transported in travelling containers that are comfortable, safe and provide adequate ventilation. Large cardboard boxes can be adapted, or travelling boxes and show cages are available from suppliers. They must be suitably placed and attached in the vehicle to avoid movement. Don't forget to take wood shavings, drinker, bottle of water, feed and the paperwork with you. On arrival at the showground go to the entrance indicated by the show secretary, for veterinary inspection. Then the bird can be placed in the show cage indicated by the stewards. Make sure that the bird has a drinker with water, and that the correct notice is attached to the cage, showing the *Breed, Class* and the *Owner Number.* After that, it's a matter of standing back and letting the bird bask in admiring glances.

Dealing with problems

You're either part of the solution or you're part of the problem .
Eldridge Cleaver, 1935

The first thing to do, if you find a sick chicken, is to separate it from the others, not only to protect it against pecking and bullying, but also to avoid possible spread of infection to the rest of the flock. Ideally, it should be put in 'hospital' accommodation, separated from, but in view of the other birds, so that if it recovers, it will be accepted by them and not treated as an outsider to be pecked. A small broody unit is ideal for this, but if there is no other shelter available, a large box on its side, with plenty of wood shavings will do. Make sure that it is secure against predators and that food and water are available, even if it is showing no inclination to feed or drink.

Sometimes a bird can be really quite sick and still make a remarkable recovery. Most vets would advise that a solitary hen be culled, but where it is part of a domestic flock or a pet, the owner is understandably reluctant to do this. Apart from providing warm, comfortable and secure conditions, what else can be done?

The following are the main problems that may be encountered, and the steps that can be taken to combat them. The home remedies that I suggest may not always work, but they can do no harm, and may do some good. I must emphasise that I am not a vet, and the advice does not replace accurate veterinary diagnosis and treatment. If a bird shows no signs of improvement after 48 hours, or if other members of the flock are affected, veterinary advice should be sought. It must be said, however, that many vets, particularly in urban areas, have little or no experience of poultry

Wounds

Any cut, graze or other wound should be cleaned with warm water and sprayed with a veterinary aerosol. The hen may need to be separated from the others in case the wound invites pecking. If it has healed over but there is obviously a gathering of pus underneath, an effective way of bringing this to a head is to apply a little *Vaseline* mixed with honey. The latter exerts an osmotic pressure, drawing the pus towards it,

A broody coop or similar can be used as hospital accommodation to isolate a sick bird and prevent the possible spread of infection.

Typical stance of sick bird

Avoiding Problems

- Buy birds that have been vaccinated against Marek's disease and Newcastle disease, and come from a flock that has been tested for Salmonella.
- Clean houses regularly and dispose of droppings.
- Move the birds to clean land regularly.
- Avoid having the flock on very wet ground.
- Avoid mixing birds of different ages unless absolutely necessary.
- Give a balanced diet of proprietary layer's feed, grain, grit, crushed oystershell.
- Ensure that they have fresh, clean water every day.
- Clean feeders and drinkers regularly.
- Check birds and houses regularly for the presence of lice and mites.
- Get to know your chickens so that any problems can be identified immediately.

until the abscess bursts. Once this has happened, clean it up as before, then apply a veterinary spray. This imparts a nasty taste to any future attacker.

Bumble foot

This is where a wound underneath the foot has healed on the surface but has pus underneath. Watch out for any birds that are limping. Treat as detailed for *Wounds*. It is important not to have perches placed too high, particularly with heavy breeds, for jumping down and landing heavily may cause the condition. Areas where the ground has sharp flints may also cause damage when the birds scratch about.

Convict's foot

If birds have been scratching about in straw or wood shavings litter, and the ground is also wet, the claws may acquire a 'wattle and daub' build up which forms a ball and sets hard. It looks rather like a convict's ball and chain being dragged around. It is impossible to pull it off without damaging or pulling away the claw, and one appreciates the long-lasting qualities of wattle and daub in vernacular housing. The best way is to soak the foot in warm water and gradually remove the debris.

Impacted crop

If unsuitable material such as long stalks of grass have been eaten, there is danger of crop impaction, in other words, a blockage. If a hen has stopped eating, looks miserable and the crop feels really solid to the touch, this could be the problem. Trickle in a teaspoonful of olive oil to lubricate the crop contents and knead the crop area gently to try and get things moving. Once contents have moved further down, they will be subjected to gastric juices that will help to break them down. In the gizzard, the strong muscular walls and small stones grind them up. Ensure that the birds have access to poultry grit. If none of this works, it may be necessary to get the affected bird to regurgitate the crop contents. Hold it with its head down, keeping the beak open with finger and thumb, and gently knead the crop, moving the contents down towards the beak. Take great care not to overdo this in case it chokes.

Sour crop

This is another condition of digestive upset, perhaps as a result of an unbalanced diet (such as a surfeit of scraps) or giving proprietary feeds which have gone musty. This time, the crop will probably feel 'squashy'. Add a teaspoonful of Epsom salts to a glass of water, then trickle in a teaspoonful of the resulting liquid into the bird's beak. For the next few days, instead of its normal diet, crush some layer's pellets and mix with a little plain, live yoghurt until the mixture looks like thick porridge. Feed this, along with a drinker of fresh water. The yoghurt has a probiotic effect in helping to replace the normal, healthy micro-flora of the digestive system. (Normal droppings are fairly firm and brown with a white surface).

Egg bound hen

This condition is comparatively rare, but easy to recognise when it does happen. A large egg has become stuck in the oviduct and the hen cannot lay it. She looks miserable, stops eating and goes in and out of the nest box to no avail. Quick action is called for! First, put *Vaseline* around the rim of the vent. Then, hold the hen *securely* above a pan of boiling water so that steam bathes the vent area and relaxes it. Take great care to ensure that she is high enough to allow the steam to cool sufficiently before it reaches her, so that there is no risk of scalding.

If these measures do not work, and the egg is low enough to be seen inside the vent, puncture it from the outside and use the fingers to 'hoik' out the shell from the vent. Take care to ensure that it all comes out for any remains can cause an infection. I heard from a poultry keeper that her local doctor had given one of his egg-bound hens some *Valium* in order to relax her. It apparently worked and the egg came out, but one wonders if the hen had other withdrawal symptoms.

Farmer's lung

This is where the lungs and windpipe are affected by intake of spores from the fungus, *Aspergillus fumigatus*. It can affect humans as well as birds and is usually the result of using damp hay in nest boxes or allowing straw litter to become compacted and wet. It may also emanate from mouldy food. Use dry wood shavings in nest boxes and replace frequently.

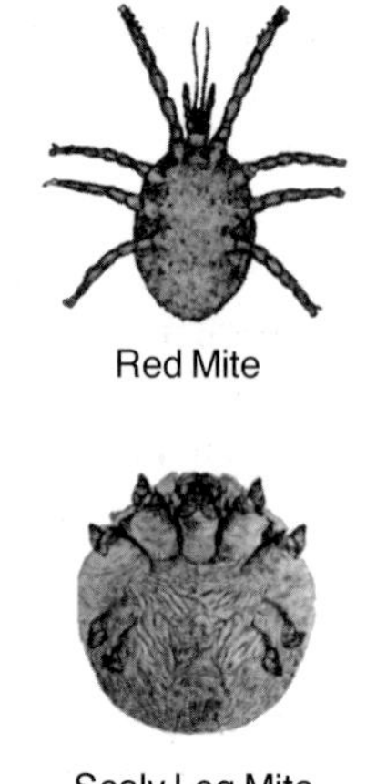

Red Mite

Scaly Leg Mite

Treating a hen for lice

External parasites

There are a number of these pests which can cause misery and even death if the infestation is large and prolonged. In warm weather a mild attack can reach epidemic proportions in a short time, so immediate action is essential.

Lice: These are greyish and are found scuttling around on the skin when the feathers are parted. They are around 3mm in length and white egg clusters are often found cemented to the base of the feathers. Giving a dusting of a *pyrethrum* based product such as *Derris* is effective, but it will have to be repeated after 7 days in order to kill off any 'nits' that have hatched subsequently. Liquid spray products are also available from vets, licensed suppliers and larger pet shops.

Fleas: Poultry fleas are sometimes found clustered like tight brown spots around the head. They often lay their eggs in the nest box litter so it is important to treat all the house as well as the birds themselves. *Derris* is effective, as are any of the spray products available from licensed suppliers.

Ticks: These nuisances are comparatively rare, tending to be found on birds that free-range in long grass. A tick attaches itself to the skin so that it can suck blood from its host. As it engorges, it swells up in size, changing from grey to a reddish colour. It is important not to try and pull it off because the head parts will be left behind, causing an infection. Dab it with some surgical spirit and it will fall off of its own accord. Alternatively, give it a quick spray with a proprietary product.

Red mite: The Red mite, *Dermanyssus gallinae,* hides during the day in crevices in the walls, roof, perches and nest boxes of the house, coming out at night to feed on the hens when they go to roost. About ½ mm in length, it is potentially one of the most serious threats to the well-being of the chickens. A bad infestation can kill because of the severe debilitation it causes. The birds become anaemic, then emaciated and finally succumb. Yet, those new to poultry keeping may not even realise that the mites are there until too late. A warning sign is when hens are reluctant to go into the house at night, particularly if they have previously done so without any problems. It is important to clear everything out of the house and treat it with a proprietary product. In severe cases, blow-torching the inside of the house is effective.

Northern fowl mite: This mite, *Ornithonyssus sylvarum,* is found on the birds during the day and so differs from the red mite. The birds and the house should be treated with an anti-mite product.

Scaly leg: This is a condition where microscopic mites, *Dermatoryctes mutans,* burrow into the skin, in between the scales of the legs. Their activities produce white crusts which push up and distort the leg scales, causing severe leg problems if not dealt with. It is an extremely infectious condition, and it is necessary not only to treat the birds but also to move them onto clean ground and thoroughly clean the house.

If there are large areas of crusts, soak the legs of the birds in warm, soapy water to dislodge the crusts. Don't try and pull them away, because this will tear the skin and cause bleeding. Dry the legs and if there has been no bleeding, paint on a proprietary product such as *Benzyl benzoate* which is available from vets and chemists. Paint on a second application a week later in case there are any eggs that have subsequently hatched.

Feather and vent pecking

Sometimes hens will attack other birds, pecking and pulling feathers, and drawing blood. If this happens, the blood-stained area will invite further attacks, and if the victim is not able to escape, she may be killed. Severe attacks can also lead to cannibalism.

An application of a veterinary aerosol spray onto the affected area will not only help to heal any wounds, but also imparts a nasty taste which helps to deter further attack. Try and identify the culprit, separate her from the rest and put her in a temporary pen next to the normal pen. This ensures that she can still be seen by others so that, when returned, she will not be attacked as a stranger. Hopefully the separation will break the habit. Next, try and determine the motive. Hens will attack new birds, as well as those that are ill or injured. It may not necessarily be the pecking order manifesting initially, but a shortage of protein in the diet. In this situation, feathers which are virtually all protein, become very attractive to make up the deficit. An outbreak of lice or mites may also trigger off feather pecking. The annual moult, where feathers are lost naturally to be replaced by new ones, may also encourage the habit. Boredom may be a cause, particularly in a small run. Hang up greens for them to peck.

Internal parasites

Leg scales pushed up by encrustations produced by the activities of scaly leg mites.

There are a number of internal worms that can affect chickens, but if hens are given access to clean land on a regular basis, they are less at risk. Don't let the grazing area become waterlogged. This encourages snails, and snails are one of the hosts of internal worm parasites. Don't throw snails into the run for them to eat, for the same reason.

The fallow ground should be raked and given a sprinkling of lime which helps to 'sweeten' the soil and deter pests. However, it is the resting period which helps to break the life cycle of pests such as internal worms, for they are unable to complete their development because their final hosts (the birds) are not there, so they die.

Warning signs to look out for in chickens when internal worm infestation may be suspected are: anaemic looking combs and wattles, rough plumage (not to be confused with the natural moult in late summer), a sharp keel (breastbone) with very little flesh on either side, and a permanently mucky bottom. The main internal parasites are *Roundworms* which are about 5cm long, *Tapeworms* which can grow to around 10cm, and *Gapeworms* (although these are usually only a problem if there are pheasants around). Gapeworms can be a problem with young birds; they attach themselves to the mucous membranes of the windpipe. They are red and around 15mm long.

Only suspect gapeworms if a bird is continually stretching out its neck, opening and closing its beak (gaping), shaking its head, while making no sound. There may also be bubbles around the beak. It may be that something has just gone down the wrong way, of course, so it's a regular pattern of behaviour that we're looking for. The traditional way of dislodging them was to take a quill feather, strip away the barbs except for the tip, then dip it in turpentine and carefully insert it into the bird's trachea and rotate it to dislodge them. This is not recommended! It is too easy to damage the windpipe. Giving a wormer is far more effective and humane.

Flubenvet is a poultry wormer that is suitable for dealing with *all* worms. It is a granular product which is added to the feed every day for a week. After that, there is a withdrawal period for table birds but not for eggs, as used to be the case. (A withdrawal period is a period of time when the bird should not be slaughtered for human consumption - usually a week). The smallest quantity of *Flubenvet* that you can buy is a pack of 240gm. This is enough for 200 kilos of feed.

Sprinkle in the required dosage and mix it up with enough feed for one week, for the whole flock. You'll have a lot of the wormer left over but it will store for quite a long time, or you can share it with other poultry keepers.

Egg eating

We've all come across this problem! When you open the nest box lid to collect the eggs a rogue chicken has got there first. Try and identify which one it is and if possible, separate her temporarily, as described earlier. A change of environment may break the habit. Collect eggs regularly: any left in the nest box provide temptation. Check that the birds are having enough food and fresh, clean water. Thirst or insufficient protein and minerals may trigger the habit. It is more common in hot weather but can occur any time.

If it is not possible to identify the culprit, try putting some 'pot' eggs in the nest box. Another ploy is to crack an egg on one side, remove the contents and replace them with hot curry paste or mustard and tape over the crack. Place the egg, tape-side down in the nest box.

Nest boxes should be in the darkest area of the house. Placing a 'curtain' of plastic strips across the entrance to the nest box is effective in making it a 'laying only' area. Eggs tend to be pecked if they are easy to see. If none of these work, consider getting some 'rollaway' nest boxes. These are sloped so that as an egg is laid, it rolls backwards and lands in a collection gulley which a probing beak can't reach.

Unwanted broodiness

Having a broody hen may be useful if you have fertile eggs to incubate, but unwanted broodiness is a nuisance. The broody sits tight on the nest box so that it is unavailable to other hens who want to use it for ordinary laying. Remove her and put her in a temporary run where she can still be seen by the rest of flock in case she is attacked as a stranger when returned. An enclosure such as that shown on page 16 is suitable or you can make your own by hammering in a few posts and placing netting around it. The secret of breaking the broodiness pattern is to cool the hen down, so don't give her a house, although rain and sun shelter are necessary. A good way of providing this is to open an umbrella and place it above the run. In this way, she has shelter from above but is open to the air from the sides. (Make sure that it is in an area where predators cannot gain access). A day or two of these conditions and she has usually forgotten about being broody, although she may try it again. Make sure that eggs are collected regularly for the sight of a clutch of eggs may trigger the habit.

Dealing with aggressive birds

Problems in this area are two-fold; hens that are aggressive towards each other and male birds that are aggressive to humans. In the first instance, follow the procedure already outlined for dealing with feather pecking.

To deal with an aggressive male, it is necessary to understand the behaviour patterns involved. If he spots a rival male (the poultry keeper may also be seen in this role), he lets out a series of sharp, staccato but rather strangled squawks, then raises his wings slightly, trembling the ends as he does so. He then adopts a series of short, sideways hops before jumping and administering a slashing blow with his spurs. If the attack is successful, he either repeats the tactic or resorts to a series of pecks, particularly if the rival stands his ground. If the latter turns and runs, he will be pursued for a short distance before the victor retires and continues his triumphant progress.

This pattern of behaviour is very similar to sexual display. When he spots a female that is 'in-lay', he utters the same sounds, followed by the sideways hops and struts, before grasping her by the feathers on the head or at the nape of the neck, and jumping on her back. A cock will not mate with a hen that is not currently producing eggs.

In dealing with a vicious cock, first make sure that small children do not have access to the run. Then, the secret is to play him at his own game! If he starts to take sideways steps and is about to jump at you, do the same. Jump heavily to the side, several times, making sure that you really stamp your feet, and hold your elbows out aggressively. If you can imitate his sounds, so much the better. (Onlookers will assume that you are eccentric, of course).

Chickens, like all non-predatory birds, have *monocular vision* (their eyes are placed on either side of the head so that they have a better all-round vision to protect them against predators). Consequently, their aggressive actions are also sideways on. You have an advantage over a cock because you can not only imitate his sideways manouvres, but can also focus your *binocular vision* (both eyes at the front) to better effect, and see precisely where he's aiming. This is usually the leg, so make sure you wear tough jeans before starting the subduing process. After a few skirmishes, you'll have the precision of a ballet dancer, and be able to launch a kick that lifts rather than hits, so as not to hurt or damage him.

Pressing him down hard on the back, is also effective, as it is for a hen who is throwing her weight around. If your performance is convincing, he'll retire, beaten. If not, try grabbing the aggressor and dunking him in a bucket of water (not in winter). The shock and affrontery to his prestige is sometimes sufficient to have him running off to a corner where he can ruffle and preen his feathers. If it's not *'a sadder and wiser cock who wakes the morrow morn'*, admit defeat and get rid of him or arm yourself with a dustbin lid!

Bacterial and viral infections

Diagnosis is the work of the vet. Bacterial infections can be treated with antibiotics but these are available only on prescription. A veterinary book is useful to have, but although it may help with diagnosis, it should be seen as an aid to knowing when to call the vet, rather than replacing his services.

Bacterial infections

Fowl typhoid - symptoms are listlessness and sulphur yellow droppings.
Infectious coryza - inflamed eyes and nose, sneezing and facial swellings.
Infectious synovitis - sneezing, swollen hocks and blisters on breast.
Fowl cholera (Pasteurellosis) - no appetite, swollen, sometimes bluish wattles.

Viral infections

If diseases are caused by viruses, there is not a great deal to be done, apart from possibly treating the side effects with antibiotics, as these are often caused by bacterial infection in the wake of the main condition. Again, it will need to be a vet-prescribed prescription. Some of the serious viral infections are avoided by vaccination or immunisations, but not all viruses have vaccines available for them. Where they are available, commercial poultry are treated at day-old or at a young stage, depending on the vaccine.

Birds can get viral infections and recover from them, of course, so that they then have antibodies to protect them against future infection. It is important, however, to keep birds of different ages separate, if the sizes of flocks are relatively large. On a small scale, where perhaps up to six or a dozen birds are kept, this may not be practical. At the same time, small flocks in healthy conditions are less likely to have the range of infections that affect birds in greater numbers.

Newcastle disease - Laboured breathing, paralysis or twisted neck.
Infectious bronchitis - Gurgling and wheezing, deformed eggs - wrinkled shells
Infectious laryngotracheitis - Coughing (sometimes blood), eye conjunctivitis.
Avian influenza - Sneezing, nasal discharge, swelling of head and neck.
Marek's disease - weight loss, paralysis (the classic form of paralysis is one leg forward, one leg back).
Fowl pox - Brown wart-like lesions on the comb and head, little appetite.
Viral arthritis (Tenosynovitus) - Reluctance to walk so bird sits on its hocks, swollen leg tendons.

Newcastle disease is a notifiable disease, and if you suspect its presence it must be notified to the authorities. This is all very well, but how is the small poultry keeper to know if a hen has Newcastle disease? Perhaps it is wheezing because it has a cold. However, if paralysis then sets in, it's important to contact a vet and report the condition.The vet will then be able to advise on the best course of action. If it proves to be *Newcastle disease,* he will inform DEFRA on your behalf If not, he may prescribe an antibiotic in the water.

Killing chickens

At some point, a bird may need to be killed. This should only be done by those experienced. A vet will do it where a sick bird is involved. Where it needs to be done in an emergency, or where birds are to be culled for the table, specialised equipment is available from suppliers.

Quirky Questions

Questions are much more interesting than answers

I write on poultry for *Country Smallholding* magazine, and also give talks to various agricultural colleges, poultry groups and schools. The following questions include those that are asked regularly, as well as some more unusual ones that have certainly taxed the brain at the time of asking.

Are some chickens more intelligent than others?
It would appear so. I have a hen that is as canny as they come. She worked out that if she gave the spacer of the feeder a quick kick, it would make the whole thing spin round so that the feed was thrown out onto the ground. This is despite the fact that the spacer is supposed to prevent feed wastage. She is always the first out of the house in the morning and the most difficult to catch if she is where she shouldn't be.

Can you use dogs to control hens?
Sheepdogs have certainly been used to control flocks of ducks, and indeed this is often demonstrated at agricultural shows. But what about hens? I had a letter from Joop van Montfoort in Porlock, to say that he used his Border Collie to keep the chickens out of the vegetable garden *(Country Smallholding Issue 80, 1989): "We've trained our hens to avoid areas like the kitchen garden by sending our dog after them every time they trespass."*

Can chickens see colour?
They certainly can, otherwise all those beautiful glowing colours in the male plumage would be wasted during his display and courtship.

How big an egg can a hen lay?
This is a bit like the 'how long is a piece of string' question. L. Pearson from Dorset wrote to me *(Issue 80, 1989)* to say that his 30 week old Arbor Acre hen had laid an egg that weighed 150g and measured 86mm in length and 178mm in circumference.

Are there safe paints for poultry?
Some poultry keepers make decorative hen houses for the garden, and then paint them with left-over paints. It's important however, not to be tempted to use old pots that might have been in the shed for a long time. They may contain lead and if the paint subsequently flakes, they will almost certainly be picked up by the birds, with possibly fatal consequences. Buy paints that are marked as being safe to use where children are concerned because what is safe for children will be alright for chickens. Similarly, if a timber house is to be proofed against damp, use a proofer that is non-toxic.

For how long does a chicken live?

It is often claimed that traditional breeds live longer than hybrids, but there is no evidence to support this. The perception has probably come about because commercially, hybrids are only kept for a relatively short period before being disposed of, and during that time are pushed to peak production. Hybrids kept on a domestic, non-intensive scale can live just as long as the old breeds. There is some evidence, however, that bantams may be more long-lived. The oldest bird I heard of was a bantam cross called Irene that succumbed at the grand old age of *15. (Country Smallholding, July 1994).*

For how long can a chicken remember?

I don't mind admitting that this question really threw me at the time, and then I remembered a letter from a reader of the magazine dating back to *Issue 93, 1991.* S.V and B.E. Taylor of Redditch wrote in to say that they had needed to confine their free-range hens in a large shed for 6 weeks. When they were then released, the Taylors reported that, *"Within a few minutes two of the hens headed across the yard, one to her old nesting place on the top shelf in the toolshed, and the other to her old place under the hedge up the drive. We checked later and both had laid."* It goes to show that while elephants never forget, chickens remember longer than we realise.

Is it true that a chicken can die of fright?

I heard from a Spanish friend that when a pair of eagles flew low over his cortijo in Southern Spain, one of his hens ran to the shade of a bush, then died. It could have been coincidence, and that the hen had some undiagnosed condition, but large birds of prey and low-flying aeroplanes can cause panic amongst free-ranging birds. They should always have outside shelter. Conversely, it seems that chickens can be extremely durable, judging by a letter I had from Claire Greenhow *(Issue 49, 1983).* She had an Ancona hen called Paxo which disappeared during a heat wave. Paxo was discovered six days later, wedged tightly between some hay bales. *"She couldn't walk, but after a few days she was fine".*

Why won't hens go in?

Newly bought birds need to be placed in the house on arrival, and kept confined for a while until 'home' is imprinted. Just to

Pressing down on the back of an aggressive bird so that she 'squats' may improve matters. Many hybrids squat automatically when approached because they are increasingly being bred for docility.

release them into a run means that they take longer to discover where to go in and perch. However, if they have normally gone in without any problems before, suspect that the house is infested with Red mite. As this pest tends to hide away in crevices during the day, they will not be apparent on the birds. They come out and attack the chickens when they go in to roost at night, so a reluctance on the part of the birds to go in is understandable.

Why won't they use the perch?

Check that it is not a problem with Red mite, as detailed above. If they are newly-bought pullets, and they have been completely floor-reared, they may not have been used to perches. If this is the reason, you will need to train them. Lift each one in turn and place her on the perch. Rescued battery hens will also need to be shown. Another reason is that the perch is unsuitable. A broom handle, for example, is too round and slippery. A perch needs to be around 4-5cm wide and slightly curved at the top around it. An 8cm x 8cm perch when bevelled at the top will provide a 4-5cm width.

How do chickens pee?

Commonly asked by primary school children, this is a glorious 'lavatory' question that would flummox most adults, if they were honest. The fact is that the urine does not emerge separately from the faeces. It travels from the kidneys, down the ureter tubes to the cloaca. Meanwhile, the faeces pass down the large intestine to the cloaca. Both are then ejected in the droppings from the same opening - the vent. The urine is the white part of the droppings. The urinary system is the same in males and females.

What is the pecking order?

The pecking order is essentially an 'order of rank' hierarchical system of dominant in relation to submissive. A cock bird or old hen may regard itself as the 'top bird'and will reinforce this by occasionally aiming a peck at the others. From the 'top bird' there is usually a chain of command extending right down to the lowliest member of the flock. This is a bird that is young, weak or a stranger, and has to spend much of its time dodging the pecks of the rest of the flock. A bird higher up the pecking order will show aggression in order to maintain its position. Similarly, a bird lower down the scale will show deference in order to escape being pecked and hurt. For a harmonious situation, the owner needs to establish himself/herself as 'top bird' within the flock structure, as referred to earlier in the book.

Do hens cooperate in rearing chicks?

It is quite common for two broody hens to share a nest, although it may not be advisable because the eggs become mixed up and may be at different stages of development. What is less common is for two hens to share the duties of looking after *all* the chicks once hatched. David E. Perril *(C. G. Issue 66, 1986)* described how one of his bantams, Bunty, was given a clutch of fertile Khaki Campbell duck eggs while another, called Emily, had a batch of White Leghorn bantam eggs. All the ducklings died and Bunty immediately went and sat by Emily's nest. (She could not get in because there was wire netting around it). As soon as the Leghorn chicks hatched and the netting was removed, Bunty moved straight in to join Emily. David wrote that: *"From that moment, the duties of raising the chicks were shared equally between the two mother hens. They strutted proudly round the garden with their shared brood. At night, both hens sat on the chicks together in the nest."*

Jenni Frost of Abergele had a similar experience, where two hens shared the same nest during the incubating, hatching and brooding periods. *(October 1998) "The first hen went broody one week before the other and sat on her six eggs. Her friend then also went broody right next door to her in the same nest box. As she laid her eggs, one by one the other hen took them into her nest and they took it in turns to sit on them. As the chicks emerged, one by one the second hen took charge of them, protecting and nurturing them until 10 had emerged and two had died. The first hen's job now complete, they both take charge over all ten during the day and, at night one of them teaches the older chicks to perch, whilst the younger ones sleep with the second hen on the floor."*

Why does the moult seem to go on for ever?

It shouldn't do. I once heard from a poultry keeper who had this problem but it turned out that he was feeding his hens on nothing but oats. This is the most 'heating' of the grains and they were being fed this ration in summer! The diet was also deficient in proteins which are the main constituent of feathers. Once on a normal ration, the birds quickly re-feathered.

Another poultry keeper revealed that his little terrier used to 'play' with one particular hen, but he couldn't understand why it seemed to be in permanent moult. The poor thing was terrified, and stress was causing the abnormal feather loss.

If the moult seems excessive or prolonged (more than a couple of weeks) ensure that their diet is a balanced one. Also check in case the birds have mites and take appropriate action.

What can be done to protect flower beds from chickens?

I use green plastic coated wire fencing of the type that is widely available in garden centres. It is decorative, with loops at the top, and does not look out of place in a garden. It doesn't have to be very high and does seem to work, probably because the hens have access to the lawn, as well as to a shady corner of soil where they take dust baths. It is interesting that just the presence of a barrier has a deterring effect; if the hens stopped to 'think' about it, they could go over with ease. A friend who lives nearby has the same experience. *"I put 2' 6" high fancy wire around the flower beds and just let the hens roam. Sometimes the odd one does get in amongst the flowers, but little damage is done if the beds are well established".*

Are chickens attracted to bright objects?

There does seem to be something of a thieving magpie instinct in chickens, if the experience of B. Bremner is anything to go by. She wrote to say that as she was bending down to look in the nest box, a chicken that was in there reached up and pulled a gold stud ear ring from her ear, then swallowed it! *"It was fortunate "*, she said, *"That it was a cheap Argos stud, and not one of Grandma 's heirlooms. The chicken is now worth £1.50 more than she was before". (August 1997 issue)*

Why do eggs vary in shape and quality?

There are many reasons for this. When a pullet first starts to lay, the eggs can be quite small, and there may even be a 'wind' egg (one without a yolk) until the system settles down. As hens get older, the shell quality declines, acquiring a more papery texture and the eggs themselves may be rather misshapen. There may also be rough patches on the shell. Sometimes a shock or a disturbance to the normal routine can cause problems in the hen's egg laying system. Low flying aircraft, birds of prey, noisy dogs, a thunderstorm, can all have their effect. An egg suddenly stops in its passage down the oviduct, acquiring an extra layer or band of shell in the process. When it recommences its journey and is subsequently laid, the elongated egg, with its extra band is testimony to the shock. I have also observed that if the birds get soaking wet, there is always one that lays an egg without a shell next day. (Chickens are not always bright enough to seek shelter when it starts to rain. Their feathers do not have much oil in them to repel water).

When two eggs reach the albumen producing area and the shell gland at the same time, the result is a double yolker. Triple yolkers are not unknown. If the chickens eat a lot of Shepherd's Purse plants or acorns, the effect is to turn the yolks green. In winter when the grass is not growing, the yolks can be very pale, gradually becoming darker as the grass grows again.

A number of diseases such as Infectious bronchitis and Egg drop syndrome can also affect egg quality. Disease or stress can produce eggs with very watery albumen, for example. This can also be caused by ammonia from badly maintained litter, or it may be a reaction to a vaccination. Only a vet can provide a definitive answer.

Is there anything we can catch from chickens?

Newcastle disease, which is a serious, notifiable disease in poultry, can manifest as a mild form of flu in humans, but it is rare. Psittacosis, which is usually referred to as parrot disease, can be transmitted to humans from chickens. This can be fatal but again is extremely rare, particularly as household chickens are not confined to indoor bird rooms where dust accumulates. Cryptosporidium, salmonella, campylobacter can all cause disease and diarrhoea in humans. Maintain a good standard of hygiene and always wash your hands after handling chickens.

Reference Section

Publications

FREE - RANGE POULTRY. 3rd edition. Katie Thear. Whittet Books 2003
INCUBATION: A GUIDE TO HATCHING AND REARING. 3rd edition. Katie Thear. Broad Leys Publishing Ltd. 1997
BRITISH POULTRY STANDARDS. 5th edition. Edited by Victoria Roberts. Blackwell/Poultry Club of Great Britain. 1997
STARTING WITH BANTAMS. David Scrivener. Broad Leys Publishing Ltd 2002
THE CHICKEN HEALTH HANDBOOK. Gail Damerow. Storey Books. USA. 1994
POULTRY OF THE WORLD. Loyl Stromberg. Silvio Mattacchione. 1996
COUNTRY SMALLHOLDING magazine. 01392 888588 www.countrysmallholding.com
SMALLHOLDER magazine. 01326 213333 www.smallholder.co.uk
PRACTICAL POULTRY magazine. 01959 541444 www.practicalpoultry.com

Housing

The Domestic Fowl Trust 01386 833083 www.domesticfowltrust.co.uk
Eggstra Special 01508 550267
Fishers Woodcraft 01302 841122 www.fisherswoodcraft.co.uk
Forsham Cottage Arks 0800 163797 www.forshamcottagearks.com
Gardencraft 01766 513036 www.gcraft.co.uk
Hodgsons 01833 650274 www.hodgsontimberbuildings.co.uk
Jim Vyse 07970 533764
Lifestyles UK Ltd 01527 880078 www.lifestylesltd.co.uk
Lindasgrove Arks 01283 761510
Littleacre Products 01543 481312 www.littleacre.com
The Poultry Pen 01673 818776 www.poultrypen.co.uk
Rivers Animal Housing 01233 822555 www.riversanimalhousing.co.uk
Smiths Sectional Buildings 0115 925 4722 www.smithssectionalbuildings.co.uk
SPR Poultry 01243 542815
V-Plas 0808 143350

Equipment

AXT-Electronic 0049 3691 721070 www.axt-electronic.de
Ascott Smallholding Supplies 0845 1306285 www.ascott.biz
Cyril Bason (Stokesay) Ltd 01588 673204/5 www.cyril-bason.co.uk
The Domestic Fowl Trust- see Housing
Hengrave Feeders Ltd 01284 704803
Parkland Products 01233 758650 www.parklandproducts.co.uk
Rooster Booster 07762 298373 www.roosterbooster.co.uk
Smallholder Supplies 01476 870070
Solway Feeders Ltd 01557 500253 www.solwayfeeders.com
Torne Valley 01302 756000

Poultry Feeds

W & H Marriage & Sons Ltd 01245 612000 www.marriagefeeds.co.uk
Small Holder Feeds 01362 822900 www.smallholderfeed.co.uk

Pure Breed Poultry Collections on view
The Domestic Fowl Trust - see Housing
The Wernlas Collection 01584 856318

Incubator Suppliers
Aliwal Incubators 01508 481729
Brinsea Products Ltd 0845 226 0120 www.brinsea.co.uk
Interhatch 0700 4628228
MS Incubators 01454 329233
Southern Aviaries 01825 830930

Electric Fencing and Netting
Electranets Ltd 01452 617841/864230 www.electranets.com
Electric Fencing Direct Ltd 01732 833976 www.electricfencing.co.uk
Hotline Renco Ltd 01626 331188 www.hotline-fencing.co.uk
G.A. and M.J. Strange 01225 891236

Organisations
Poultry Club of Great Britain 01205 724081 www.poultryclub.org
Rare Breeds Survival Trust 024 7669 6551 www.rare-breeds.com
Rare Poultry Society 01162 593730 evenings
Utility Poultry Breeders' Association 01926 420962 evenings www.utilitypoultry.co.uk

Buying Chickens
There isn't room here to list breeders and suppliers, but there is a regularly updated Breeders' Directory in every issue of Country Smallholding magazine. It also appears on their website at www.countrysmallholding.com

The author with a Speckledy hen.

Chickens are adapted for outside living, not cages!

Index